Global Climate Change

Global Climate Change

A Guide for Future Action

Malcolm J. Prowle

Global Climate Change: A Guide for Future Action

Cover design by Brent Beckley

Interior design by Exeter Premedia Services Private Ltd., Chennai, India

First published in 2024 by
Business Expert Press, LLC
222 East 46th Street, New York, NY 10017
www.businessexpertpress.com

ISBN-13: 978-1-63742-724-8 (paperback)
ISBN-13: 978-1-63742-725-5 (e-book)

Business Expert Press Environmental and Social Sustainability for Business Advantage Collection

First edition: 2024

10 9 8 7 6 5 4 3 2 1

Description

Recently, large parts of the world faced record high temperatures–another item on a long list of warning signs about the climate. This book is for the reader interested in having a better understanding of the nature and causes of climate change and the measures being undertaken.

Global Climate Change suggests success is unlikely, and to prevent catastrophe, more radical and uncomfortable actions will be needed. Many groups in society (e.g., governments, economies, businesses, and public services) will need to undertake such actions to mitigate climate change and deal with its consequences. This book provides guidance about what these actions will involve.

Contents

Key Environmental Quotes

"*We are the* ***first generation*** *to feel the effect of climate change and the* ***last generation*** *who can do something about it.*"—**Barak Obama**

"***We are running the most dangerous experiment in history right now,*** *which is to see how much carbon dioxide the atmosphere can handle before there is an environmental catastrophe.*"—**Elon Musk**

"*We do not inherit the Earth from our ancestors,* ***we borrow it*** *from our children.*"—**Native American Proverb**

"*The Earth lies polluted under its inhabitants; for they have transgressed laws, violated the statutes, broken the everlasting covenant. Therefore, a curse devours the Earth, and its inhabitants suffer for their guilt.*"—**Isaiah, The Holy Bible**

"*It is baffling, that in our modern world we have such blind trust in science and technology that we all accept what science tells us about everything—until, that is, it comes to climate science. All of a sudden, and with a barrage of sheer intimidation, we are told by powerful groups of deniers that the scientists are wrong, and we must abandon all our faith in so much overwhelming scientific evidence.*"—**Prince Charles, now King Charles III**

"*Though every nation must do its part to address climate change, developed nations are responsible for the lion's share of carbon pollution in the atmosphere, and they have an obligation to help developing nations transition to a sustainable future.*"—**Frances Beinecke**

"*Corruption has spread on land and sea as a result of what people's hands have done, so that Allah may cause them to taste the consequences of some of their deeds and perhaps they might return to the Right Path.*"—**Holy Quran chapter 30, verse 41**

Preface

In July 2023, large parts of the world faced record high temperatures, something which is damaging to health. Although it is not definitive that these extreme weather events are linked to climate change, it does look very much like it.

For a long time, like many people, I listened to the various debates, comments, and warnings that were being made by climate scientists about trends in global warming and climate change. I can also remember famous people like David Bellamy, Sir David Attenborough, and Prince Charles (now King Charles III) talking and writing, decades ago, about the dangers of climate change and environmental damage. While it was interesting to listen, I cannot say that I took a deep interest in the topic.

However, some years ago, I seemed to have had a Eureka moment when it dawned on me what climate change was about and the huge dangers it posed to humanity and the Earth as a planet.

In recent years, while studying climate change, I became increasingly amazed that in spite of the dire warnings from the climate scientists, the degree of interest and response from the general public was not as high as I would have expected. In 2023, a public opinion poll in the UK found that one-quarter of those polled still thought climate change was not a serious global threat, but three in four thought it was a serious threat, with half of those (who thought it was a serious threat) thought there was still time to resolve it. Polls of voter opinion about key issues for governments often place climate change behind other more immediate matters.

Also, what struck me was that in spite of the huge focus placed on such things as the annual COP climate summits and government policy announcements on various climate change issues and so on, the level of real activity seemed nowhere near what was needed to stave off a climate disaster. Hence, this book.

What I have tried to do in this book is seven main things:

1. To provide an understandable guide for the interested reader about the nature of climate change and the reasons why it is happening.
2. To address (and reject) the claims of climate change skeptics and deniers which, being fundamentally based on vested interests, are affecting the scope of action needed to address the problems.
3. To consider the actions, regarding climate change which have been taken (and promised) by governments around the world, and to evaluate whether these are likely to be sufficient to solve the climate change problem. My short answer is a resounding no.
4. To describe what broad actions really need to be taken to mitigate climate change to a satisfactory extent.
5. To consider the implications for the planet and humanity of inadequate responses to mitigating and adapting to climate change.
6. To consider the implications for key sectors in society and possible actions that might be taken.
7. To emphasize the drastic implications for the developing world.

Dealing with climate change will be difficult, worrying and possibly traumatic for people in rich developed countries. However, this will be dwarfed by the impact on poor developing countries for whom the consequences will be catastrophic.

Although the book discusses the scientific knowledge concerning climate change, it is not a scientific book. The science is a very important issue, but climate change is fundamentally a political issue with a strong dimension of economics. At the end of the day, the actions of countries will be driven by the political and economic impacts on their populations. Basically, I will argue that the political systems around the world are failing humanity with regard to this issue.

My thanks are due to a number of people who have assisted me with the drafting of various parts of the book but needless to say, any errors in the book are the sole responsibility of the author.

Malcolm J. Prowle

May 2024

CHAPTER 1

Introduction

Background

Climate change is, undoubtedly, the biggest challenge facing humanity throughout the whole of its history. It is, largely, a consequence of the rising temperatures on Earth, both on land and sea, resulting from the discharge of high volumes of what are termed greenhouse gases (GHG) into the Earth's atmosphere. These act as a "blanket" around the planet, preventing heat from escaping into space and causing temperatures to rise. Already we are observing some of the consequences of global warming in terms of record elevated temperatures, extreme flooding, melting of glaciers, and so on. We can be certain that there is worse to come.

This is yet another book on the issue of climate change on Earth but with a different standpoint to many other books. I believe that available evidence suggests that the current policies and actions of countries around the globe will NOT deliver what is needed to counter the worst aspects of climate change, with severe consequences for humanity and the planet, particularly in developing countries. I will argue that (for good reasons) politicians are not being totally honest with their people about the radical nature of the actions and changes in behavior that are needed to resolve the climate change issue. It is insufficient just to give people the impression that all they need to do is to switch to an electric car, recycle their plastic bottles, and eat a bit more vegan food.

The actions really needed are more radical and would revolutionize our lives and our societies. In particular, they will require people in richer countries to address their consumption habits, whether that be food, energy, consumable goods, and so on. Also, different sectors in our society often have contributions they should make to counter the impacts of climate change.

The Challenge of Climate Change

Hardly a day goes by now without there being some news report or research study having something to do with climate change. In electoral terms, climate change is a big concern for a large proportion of voters in many countries (especially younger people), but perhaps not such a concern as more immediate issues like the economy. Moreover, whereas in the past, climate change was something expressed through the data of scientists and the findings of their research, in recent years, people are now seeing the impacts of climate change in front of their eyes.

The key issue with climate change is, of course, not whether it is happening (it is) but what is causing it to happen and what we can do about it. As we will see later, there is abundant scientific evidence that the temperature of the Earth is rising, and this is having an effect on the climate and will continue to do so if we don't act. Throughout its existence, the climate of the Earth has changed dramatically on a cyclical basis. The question is whether this is what is happening now or is it something different?

Over the last couple of centuries, that situation has changed, and the activities of humanity are now seen as having a significant impact on the Earth's climate. Geologists now refer to the current time period (dating back perhaps 200 years or so) as the Anthropocene geological era, this being viewed as the period during which human activity has been the dominant influence on climate and the environment. This period broadly aligns with the start of the Industrial Revolution, which resulted in the burning of high levels of fossil fuels such as coal, gas, and so on.

This is a most controversial issue with views either way, but the strong scientific consensus is now that the Earth's changing climate is a consequence of human activities. Clearly, humans can amend their behaviors and activities in such a way that will reduce the rate of climate change, but they can do little to control external factors such as volcanoes, solar activity, and so on.

The issue of climate change is clearly the issue of the current age and is probably the greatest challenge that has faced humanity during the tens of thousands of years of its existence. The issue has led to feverish international discussions by governments across the globe, with 28 world climate summits having taken place.

These events always end with communiques and commitments by the various countries, but the big issue is, of course, whether individual governments and the international community will live up to the commitments agreed at the various COP conferences. Also, whether their actions, based on these commitments, will have the effect of mitigating, at least, the worst impacts of climate change. Based on the available evidence, I have huge doubts about this. I am quite clear in my mind that the world will not achieve its two totemic objectives of keeping global temperature rises below 1.5°C and achieving net-zero emissions of greenhouse gases (GHG) by 2050.

At some time in the future, the politicians of the world will accept this and move on. In the light of this, a series of radical measures to deal with the effects of climate change will need to be considered and implemented to avoid making the situation even more difficult.

In particular, it needs to be mentioned that actions to address the challenges of climate change can be categorized into twofold as follows:

- **Mitigation:** These are actions designed to slow, and ultimately, stop global warming taking place.
- **Adaptation:** These are actions designed to deal with the impacts of global warming, and an example here would be shoring up costal defenses to prevent rising seas from flooding the land.

So far, the bulk of the activity by countries and the international community has focused on mitigation, but for the reasons I have mentioned above, it appears that the mitigation actions will be too little, too late, and so there will be severe impacts of global warming. Hence, increasing focus will need to be placed on adaptation.

The Overall Global Context

As I write this book at the start of 2024, it is informative to try and summarize the global context in which the problems of climate change exist. I do not want to go into this in great depth, and a series of bullet points should suffice, the main ones of which are as follows:

- **Superpower competition:** For many years, following the end of the Soviet Union in 1989, the USA was the sole global superpower. However, this situation has changed, and China must now be regarded as a superpower in political, economic, and military terms. The veteran diplomat and former US Secretary of State, Dr. Henry Kissinger, believes that in the current situation, neither side has little margin of political concession, and any disturbance of the equilibrium can lead to catastrophic consequences.
- **Ideological conflicts:** Winston Churchill once made a statement which implied that democracy was the worst form of government except for every other form that had been tried (International Churchill Society 2016). In 1992, political scientist Francis Fukuyama (Fukuyama 1992) published a book entitled *The End of History and the Last Man*. In the book, the author argued that humanity has reached "not just ... the passing of a particular period of post-war history, but the end of history as such." That is, the endpoint of mankind's ideological evolution and the universalization of Western liberal democracy was the final form of human government. Clearly, this conclusion was, to say the least, premature as admitted by the author himself, and we have seen growing autocracy of governments in many parts of the world. Adding to this, in many countries such as the USA, the EU, the Middle East, and so on, we see a growth in political extremism which may be of the political right, left, or based on religion.
- **Major wars:** Until the start of 2022, the conventional wisdom was that a major war in Europe would never occur again. Indeed, European armies had been making plans to reduce the numbers of battle tanks required for a European conflict. Russia's invasion of the Ukraine, and the ongoing war, has changed all that, and the final outcomes of this war can only be imagined. A more recent conflict is that between Israel and Hamas in the Middle East with, again, lack of clarity about the outcome. Besides the loss of life and destruction, these wars will have

major implications for the global economy and the economies of individual countries.

- **Global economy:** The Great Recession is the term used to describe the economic recession that affected many parts of the world between 2007 and 2008. The scale and timing of the recession varied from country to country, but the International Monetary Fund (Claessens and Kose n.d.) concluded that it was the most severe economic and financial meltdown since the Great Depression. One result was a serious disruption of normal international relations. Since the Great Recession (and the associated global financial crisis), world output has grown moderately, yet the path of economic recovery has been fragile and uneven. Recently, the IMF published a report which suggested that global economic activity is experiencing a broad-based and sharper-than-expected slowdown, with inflation higher than seen in several decades. Global growth is forecast to slow from 6.0 percent in 2021 to 3.6 percent in 2022, 3.1 percent in 2023, and 3.2 percent in 2025. This is the weakest growth profile since 2001, except for the global financial crisis and the acute phase of the COVID-19 pandemic (OECD 2023). It is indeed possible that we are in a post growth era.

- **Public finances:** This is a fiscal time-bomb caused by governments of all persuasions who continue to keep borrowing large sums of money to finance growing public services rather than via tax revenues from current generations. In 2019, the accrued global public debt was estimated at $69 trillion ($69 million million), with the bulk of that debt belonging to the governments of the USA, China, and Japan (Desjardin 2019). Because of the Covid pandemic, this amount increased sharply. This debt will be left to future generations to resolve, although they may be poorer than current generations.
- **Artificial intelligence:** Much is written these days about the benefits and dangers of artificial intelligence. Governments around the world seem to have increasing concerns about the

dangers of AI in many areas. On the dangers of AI in relation to national security, Dr. Kissinger, in the same interview, states a belief that AI will become a major factor in national security in the next 5 years.

- **Ageing populations:** The phenomenon of an ageing population consequent on growing numbers of elderly and falling birth rates is well known. To some degree or another, most countries in the world face the phenomenon of ageing populations. This has major social and economic implications particularly for public services.
- **Country instability:** In many parts of the world, there is considerable instability and insecurity. The Global Peace Index (2022) stated that "Peacefulness" has declined to its lowest level in 15 years, fueled by post Covid economic uncertainty and the Ukraine conflict. Two of the five countries with the largest deteriorations in peacefulness were Russia and the Ukraine; they were joined by Guinea, Burkina Faso, and Haiti. All these deteriorations were due to ongoing conflict. Other states where violence is high include Afghanistan, Iraq, Syria, and Yemen.
- **Future pandemics:** The Covid pandemic lasted almost 3 years and caused huge loss of life, disruption, and economic damage. The Covid virus still circulates in many countries but now seems under control. However, scientists always warn that conditions in many parts of the world are such that a similar viral-based pandemic is always possible and must be guarded against.

Hence, it can be seen that governments around the world have a number of other significant challenges as well as that of climate change. Given that the task of overcoming global climate change requires international cooperation, the above situations do not foster such cooperation.

The Fundamental Drivers of Climate Change

There is so much discussion about the factors that are driving climate change that this can cause great confusion because of the numbers of factors presented. This is further complicated by the vast range of solutions being put forward to either mitigate climate change or adapt to its impacts. For simplicity and as a means of focusing on what is important, I would like to suggest that the main and fundamental drivers of climate changes can be summarized threefold as outlined below.

Population

Basically, the more people there are on Earth, the greater will be the need for the basic commodities of food, energy, housing, and so on. As we will see later, the global population has mushroomed enormously during the 20th and 21st centuries and is still growing. With this has come a huge growth in the demands for the commodities described above. However, in saying this, we must recognize that the vast bulk of humanity lives in poverty, as illustrated in Table 1.1.

It can be seen that (given the absence of data for one large region) there must be well over two billion people living in varying degrees of poverty, with half a billion in extreme poverty. In these circumstances, the demands of such people on the world's resources would be relatively small individually. However, it doesn't need a great deal of imagination to see that, given the large numbers of humans involved, any improvements in alleviating this poverty would result in an enormous rise in demand for the Earth's resources.

Consumerism

The demands from individuals in the affluent world for commodities go well beyond basic needs for survival. Modern life means that individuals, in developed countries, have the opportunity to acquire a wide range of goods and services. This has been the case in the West for decades, but now we see the rise in consumerism in other parts of the world, such as China. Production of these goods and services requires high volumes of raw materials, energy, and so on. Furthermore, it should be noted that in some cases, overconsumption of certain commodities in affluent countries can have negative consequences. An example here is

Table 1.1. Poverty in the world

	Numbers living on <$1.90 a day	**Numbers living on <$3.20 a day**	**Numbers living on <$5.50 a day**
	Millions	Millions	Millions
East Asia and Pacific	25	154	558
Europe and Central Asia	5	20	52
Latin America and Caribbean	25	63	151
Middle East and North Africa	27	76	170
South Asia	n/a	n/a	n/a
East and Southern Africa	279	442	558
Western and Central Africa	137	263	362
Rest of World	7	9	14
Total	505	1027	1865

Source: https://blogs.worldbank.org/opendata/april-2022-global-poverty-update-world-bank.

overeating, which leads to obesity and bad health. Also, and particularly in the developed world, large amounts of resources are wasted. Take, for example, the large amounts of food that are not eaten but are thrown away. Finally, as discussed in the previous section, even minor improvements in the standards of living of the very poor would also increase demands for such goods and services unless it was compensated for by reductions in the rich world.

Capitalism and Growth

In the developed world, the capitalist economic system has been predominant for several centuries, and its influence is also growing in other countries such as China. In pursuing its aims of profitability and growth, companies aggressively market their products and also place a strong focus on innovative new products that they encourage people to buy. The existence of these marketing pressures exacerbates the levels of consumption of goods and services, which often exceed basic needs.

We have only one planet to live on. Of the nearest planets to Earth, Venus is uninhabitable because of extremely high temperatures, while

Mars presents a completely hostile environment to human life, combining extreme cold with an unbreathable atmosphere and intense radiation. Beyond our solar system, a planet orbiting a star named Epsilon Eridani, located 10.5 light-years away, is the closest known exoplanet. At the moment, we have no means of reaching other planets within or beyond our solar system; it remains a fantasy to think otherwise.

Hence, we need to care for the Earth, which is something we are just not doing.

To put it bluntly, the solution to climate change comes down to a choice between two things.

- Less people and/or
- Less consumption per individual

Many people when reading this start to feel uncomfortable, particularly about the issue of reducing global populations. However, the reality is that these are the choices to be made.

Many secularists take the view that as humans, we already have all the tools needed to solve the many problems found on the planet, including climate change, poverty, war, and so on. They are, of course, absolutely correct. Science and technology have revolutionized our lives in terms of sickness, longevity, affluence, choice, comfort, and so on. However, as we will see later, we live in a world where violence, early death, starvation, and poverty are the norm for a large proportion of humanity. While we do have all the tools at our disposal to improve life on earth, we don't use them. Why not? Well, we (including me) are often too immature, greedy, self-centered, and so on to use them, because of the impact it will have on us. In saying this, I recognize that some are more flawed than others, but we are all born with flaws that limit what we are prepared to do. Is it really likely that the population of developed countries will make the sacrifices necessary to alleviate climate change if it radically affects their current lifestyles? I think not. Politicians know this, and in later chapters, we will examine whether the rulers of many countries on Earth will really be prepared to make the necessary changes against the will of their populations.

Some might wonder whether there is any need to take on board the three issues I have discussed above, with all the difficulties involved. Many will take the view that there must be an easier way by putting our faith in potential scientific and technological developments in the pipeline that will solve the problem. As I will discuss later in the book, there are indeed such technological solutions which might come to fruition, but the problem is that they will probably be too late to solve the climate problem, and indeed they may never happen. I don't think it is a good idea to place the future of humanity on such a gamble.

Thesis of the Book

In this book, I will set out my view that the available scientific evidence suggests the climate change pledges made by governments around the world will not be adequate to stop climate change. Indeed, some of the pledges made may already be inadequate. For example, whereas most governments pledged to achieve net zero carbon emissions by 2050, India has already only promised to cut its emissions to net zero by the year 2070, which is half a century away. Given that example, might other countries decide to backtrack on the 2050 commitment?

If my view is correct, the next issue to consider is what happens if the global climate change policies are not fully implemented or fail to have the desired affects. Based on what we know, it appears that the world will face some rather frightening consequences, including

- Mass flooding of parts of countries (e.g., Bangladesh, Maldives) and, if things got really bad, of some major cities including Miami, New York, London, Osaka, and Guangzhou
- Mass migration from areas in Africa, Asia, and South America which are no longer habitable for humans
- Harvest failures and food shortages
- Energy shortages
- Military conflicts over resources such as water, food, energy, and so on

It is difficult to make precise forecasts about the magnitude and timing of the above consequences, but they are clearly serious. There can be no doubt that, at this point, governments around the world will panic and try to allocate huge amounts of resources to deal with these problems. In doing this, they may also be drawn to the use of unproven technologies such as geo-engineering methods or cloud seeding, the consequences of which are uncertain. In addition, as already noted, governments will also have to allocate huge amounts of resources in order to alleviate the consequences of climate change (i.e., adaptation).

In light of the above, it will be incumbent on a range of organizations and individuals in all countries to take actions to attempt to mitigate, and adapt to, climate change in their countries.

Structure of the Book

The book comprises the following chapters.

- **Chapter 2:** This provides an overview of the Earth as a planet, incorporating its history, physical structure, geography, biology, and so on. It is important to understand the background to current climate change.
- **Chapter 3:** This discusses the nature of climate change currently taking place on the Earth and the causal factors involved.
- **Chapter 4:** This addresses the important question of the extent to which climate change is natural as opposed to consequence of human actions. It considers the views of climate sceptics/deniers and discusses why these views are incorrect and dangerous.
- **Chapter 5:** This concerns other phenomena that are closely linked to climate change, such as the loss of diversity in nature, the destruction of natural habitats, and the exploitation of the Earth's nonrenewable resources.
- **Chapter 6:** This discusses the proposals and actions already being proposed or undertaken across the globe to mitigate the impacts of climate change or adapt to it. The key questions here are whether these plans and actions will be implemented and whether they will be sufficient.

- **Chapter 7:** This considers the factors that will influence or inhibit various countries in addressing climate change. This chapter provides an analysis of key features of a selected number of countries (who are the main culprits regarding climate change) that are pivotal in mitigating climate change.
- **Chapter 8:** This chapter outlines what really needs to happen to deal with the problems of climate change. It will go beyond what is considered in Chapter 6 and (to pick up a point made earlier) consider what needs to happen to prevent large-scale climate change and disaster.
- **Chapter 9:** This chapter considers likely future scenarios concerning climate change and covers likely impacts and possible solutions that might be implemented, some of which will be seen as very radical and risky.
- **Chapter 10:** In light of Chapter 9, this chapter aims to outline the implications of ongoing climate change for a number of key organizations/sectors, including governments, businesses, and so on, and their potential responses.
- **Chapter 11:** This chapter provides some concluding comments.

CHAPTER 2

The Nature of Planet Earth

Introduction

Scientists believe that the known universe was created some 13.7 billion years ago as a consequence of what is termed the "Big Bang". According to the Big Bang theory, all matter and all space were originally part of an infinitesimally small point called the Singularity or Primeval atom. The theory says nothing about where that singularity came from, and there seems to be no real scientific answer other than that it was some random quantum event or the result of some divine intervention.

This singularity suddenly expanded (and has been expanding ever since), spewing out all matter and energy into a vast void. As the universe expanded, it cooled down, and as it cooled, the elements started to form, stars were born from the simplest elements, and the stars formed galaxies. Therefore, the explosion of a tiny seed of matter and energy formed the universe that we know today.

The most common misconception about the Big Bang theory is the idea that matter in the singularity exploded and spread out into empty space. This is not what the theory says, and it is believed that prior to the Big Bang, space itself was also small. It is, therefore, the space itself that stretched to its present size as a consequence of the Big Bang. "Thus a better term might be the Big Stretch."

The planet we call Earth is the third planet from our sun in what we call our solar system. It lies some 96 million miles from the sun. Our sun is one of billions of stars in our galaxy, which is named the Milky Way. In turn, the Milky Way is one of billions of galaxies in the known universe. As far as we know, Earth is the only astronomical object to harbor life. Some say that the trillions of stars in the universe must mean that there are a huge number of planets and many of these planets could contain life. That may be so, but others argue that the creation of life on

Earth (let alone intelligent human life) was an event that was unlikely to happen anywhere else. The reality is that we don't know, and we probably will never know.

Before we start talking about the Earth's climate, we need to know a bit more about the Earth itself. The Earth is a huge and diverse subject, and we can only touch on the essence of just a few themes. These are as follows:

- Birth and development of the Earth
- Structure of the Earth
- Life and humanity on Earth
- Natural resources of the Earth
- The Earth's climate

Birth and Development of the Earth

Scientists believe that the Earth (and its moon) formed around the same time as the rest of the solar system, which was about 4.5 billion years ago. A dense cloud of gas, which was compressed due to gravity, grew immensely hot and heavy at its center. This eventually became our Sun, but the matter on the outskirts of this gas cloud was pushed into space due to the force of solar winds. Over time, this matter coalesced into what are called protoplanets. The Earth became the third protoplanet from the Sun.

We know today that the planet Earth is a spherical entity, with some flattening at the poles, but this wasn't always the case. Although, in the 3rd century BC, Hellenistic astronomy established the roughly sphericals hape of Earth as a physical fact and calculated the Earth's circumference. Strangely, for centuries many ancient and modern cultures persisted with a belief that the Earth was flat until modern science and pictures from space proved this to be false.

At this time, the Earth was chiefly a molten material and would have been impacted by gravitational forces, so it resembled a ball of lava floating in space. As the Earth cooled from its molten state, minerals started to crystallize, resulting in a separation. This caused the creation of the three parts of the planet termed as crust, mantle, and core.

Other events were also taking place at this time. It is believed that during the early formation of Earth, asteroids were continuously bombarding the planet, and there is speculation that they may have been carrying with them an important source of water needed for the creation of life. It also seems that when the asteroids hit the surface of the Earth at great speed, they shattered, leaving behind fragments of rock. Some speculate that nearly 30 percent of the water contained initially in the asteroids would have remained in the fragmented sections of rock on Earth, even after impact. However, this is pure speculation, which is probably incorrect, and the origin of large volumes of water on the Earth remains a great mystery, which will probably never be resolved.

Earth's early atmosphere was most likely composed of hydrogen and helium, but as the crust began to form, volcanic eruptions occurred. These volcanoes pumped gases such as ammonia and carbon dioxide into the atmosphere around the Earth. Slowly, the oceans began to take shape, and eventually, primitive life evolved in those oceans.

Structure of the Earth

As noted, over a long period of time, the new Earth cooled into a solid planet. The structure of the planet is shown in Figure 2.1.

The inner core of the Earth is believed to be a solid metallic ball made mainly of iron. The outer core, which sits between the inner core and the mantle, is believed to be a magma-type fluid layer composed mostly of iron and nickel. It surrounds the inner core and fulfills a very important role in that it creates Earth's magnetic field.

The mantle of the Earth is a layer of silicate rock situated between the crust and the outer core. It is some 1800 miles thick and comprises 67 percent of the mass of Earth.

The Earth's crust is the layer that makes up the Earth's surface, and it lies on top of the mantle. This hard rocky crust formed as lava cooled about 4.5 billion years ago. It is the thinnest layer of the Earth and has an average thickness of about 18 miles below land and around 6 miles below the oceans.

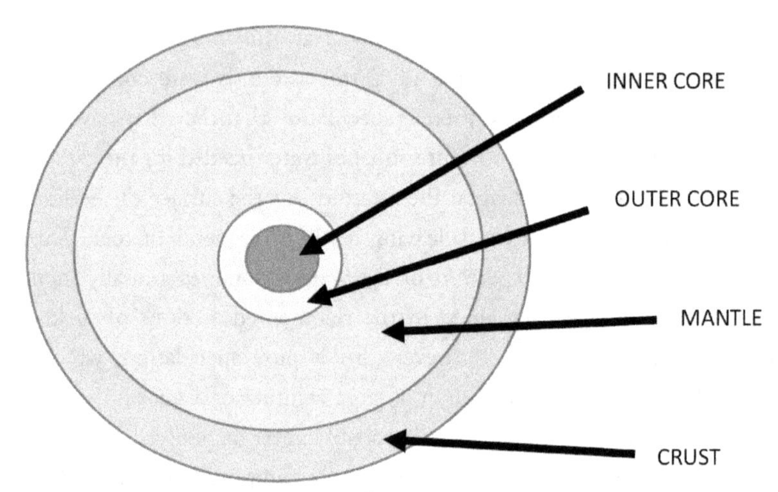

Figure 2.1 Structure of the Earth

Most of the Earth's surface, about 70 percent, is covered with water, but as already noted, the origin of that water is a mystery since the Earth started as a very hot protoplanet and all water would have been lost. The remaining 30 percent of the Earth's surface is made up of the seven continental landmasses.

The crust is broken into many large plates (tectonic plates) that move slowly relative to each other. The mountain ranges around the world formed when two plates collided, and their edges were forced up. Many other surface features are the result of the movement of these tectonic plates. The plates move anywhere from 1 to 3 inches per year. The tectonic plates are actually floating on the molten lower mantle of the Earth. Earthquakes, volcanoes, mountains, and oceanic trench formation occur along plate boundaries. The plates are in constant motion as their movement is dictated by heat dissipation from the Earth's mantle.

Life and Humanity on Earth

We use the term "life" on a daily basis without realizing that it is quite a difficult thing to define precisely as there are many definitions including those shown in Figure 2.2.

It can be seen that these various descriptions break into two broad groups—those who see life as something purely material and driven

Life, living matter and, as such, matter that shows certain attributes that include responsiveness, growth, metabolism, energy transformation, and reproduction. (Encyclopaedia Britannica)
Life is a quality that distinguishes matter that has biological processes, such as signalling and self-sustaining processes, from matter that does not, and is defined by the capacity for growth, reaction to stimuli, metabolism, energy transformation, and reproduction. (National Library of Medicine)
It is chemical in essence; terran living systems contain molecular species that undergo chemical transformations (metabolism) under the direction of molecules (enzyme catalysts) whose structures are inherited, and heritable information is itself carried by molecules. (NASA)
If we were merely physical beings, subjected to the laws of nature like everything else in the material world, then we could not act for moral reasons and hence would be unimportant. (Immanuel Kant)

Figure 2.2 Definitions of life

by the scientific laws of physics and chemistry, and those who see life (especially human life) as more than this and having a spiritual or religious dimension. The problem with the purely material approach is the difficulty of explaining how something like consciousness in a human being can be explained as an electro-biochemical entity (Chalmers 2016).

However, when we talk about life, we are not just thinking about humans or other animals, be they fish, birds, reptiles, and so on. We have to remember that plants and insects are also forms of life, and beyond that, there are all sorts of microscopic creatures who live on land and in the oceans—some of which are single-celled and others more complex. Bacteria meet the definition of life, but viruses do not.

In this section, I am going to provide a brief description of the development of life on Earth but then focus on human life and how that has grown in numbers and spread over the planet. During its history, the planet Earth has been home to over four billion different species of organisms (Raup 1986), although this is very speculative. Of all those living life forms, an estimated 99 percent of all species are now extinct.

The timeline of life on Earth begins over 4.5 billion years ago, and in the beginning, survival was difficult for any life forms. The earliest fossil evidence of life on Earth dates to 3.7 billion years ago. These were found preserved in ancient rock samples from Greenland. The fossils were once part of a now-extinct seabed and thought to be the remains of ancient microbes.

So how did life on Earth first evolve? An early proposal concerned the "primordial soup" on Earth containing water, hydrogen, methane, ammonia, and carbon dioxide. When the Earth was "young," the oceans were filled with these simple chemicals important for life. The assumption then was that somehow these simple chemicals would eventually self-assemble into the contents of simple living cells. Now this is a big jump, and it requires huge assumptions. These are that somehow or another, these simple chemicals got transmuted into the complex chemistry of life involving immensely complex chemical entities such as nucleic acids (DNA and RNA) (responsible for replication) and protein and amino acids (responsible for the creation of the physical infrastructure of the organism). I suggest this is one more of these questions to which we will never get an answer.

Subsequently, it is believed that around 900 million years ago, these simple organisms started to evolve into multicelled organisms and, in turn, would evolve into all forms of life on Earth today.

Around 540 million years ago, a mysterious event occurred. Suddenly, and seemingly out of nowhere, large numbers of species started appearing. This period is known as the Cambrian explosion and was the first time animals with skeletal systems lived. On a scale of evolution, this period in the timeline of life on Earth was by far the most exciting and vital. Soon after, the first vertebrates appear. Then, the emergence of life onto land took place rather than being confined to the oceans. Corals, ancient shelled organisms, molluscs, and soft-bodied life forms all began developing at this time.

After the Cambrian explosion, life on Earth became more diverse. Many new species of plants and animals were separated on to a path of evolution. New forms of life began to appear. Plants began colonizing the land, and fish began swimming in the seas. The first life on land started as algae gradually adapting to being able to live on dry land.

Around 400 million years ago, the first four-legged animals started to develop. These animals, known as tetrapods, were the ancestors of all birds, mammals, reptiles, and even amphibians. The first amphibians appeared soon after, living on sea floors and in shallow marine ecologies. They were the first species to branch off from the tetrapods.

Insects are next, although they took some time to develop wings. Then, reptiles developed in swamps and tropical forests, which were mainly made up of ferns and other ancient coniferous plants.

About 250 million years ago marked the first emergence of dinosaurs. The dinosaurs weren't initially the dominant species on the planet, and it would take 50 million years for them to take over the entire planet. They were pre-empted by a mass extinction event that shattered all life on Earth. Around a third of life on planet Earth was wiped out, marking a huge fork in the timeline of life on Earth. Yet, as life has shown throughout history, it continued on living.

In the late Triassic, 200 million years ago, the first mammals appeared. It was soon after the dinosaurs became the dominant species on Earth. These mammals were very small, many no larger than mice. Here, the first warm-blooded mammals appear in the record. In the timeline of life on Earth, mammals began splitting into the four major groups that can be observed today. It's not known why it happened.

It is unknown exactly when, but around this era saw the emergence of the first birds but they are thought to have split from dinosaurs or reptiles. How and when is still a mystery.

While all of this was unfolding in the animal kingdom, plant species across the world were diversifying rapidly. The first signs of flowering plants are linked to this time as the evolution of the planet flourished.

Sixty-six million years ago saw the Cretaceous-Tertiary extinction event. It was the mass extinction that saw dinosaurs wiped from the face of the Earth. It meant that mammals, which were previously living in the shadow of the dinosaurs, could take over and dominate the land. They started small, but quickly grew and diversified, enjoying their new dinosaur-free world. It was around this time that primates began spreading and diverging into new species. Lemurs, gibbons, bonobos, orangutans, gorillas, and even modern humans are all linked to this point in the timeline of life on Earth.

Around 50 million years ago, whales evolved to live under the sea. The earliest forms of these are thought to have been land animals adapting to life in the sea. They began by returning to the land only to give birth.

Over the next few million years, evolution continued among mammals. For reasons unknown, many species started getting bigger and bigger, some growing to enormous sizes. These animals, known as megafuna, populated the planet up until around 10,000 years ago when unknown events wiped most of them from the face of the Earth. Some incredible megafuna species include mammoths, woolly rhinoceros, giant beavers, and so on.

Today, hominids (humans) are the dominant species on the planet and are currently sharing the Earth with about 12 million other species. Scientists believe that human precursors had been walking the planet for about 6 million years. *Ardipithecus,* the earliest link to humanity dating 6 million years ago, was the first to walk upright. He was able to use his hands for weaponry, toolmaking, and survival needs. Between 2 and 4 million years ago, the *Australopithecus* group appeared. This group was more advanced than the *Ardipithecus:* they could walk upright, create tools, and climb trees.

Hominids (humans) appeared in Africa around 2 million years ago and began spreading throughout the world. Over a period of several millions of years, there were several species of hominids:

- Homogeorgicus: 1.8 million years ago
- Homoerectus: uncertain–50,000 years ago
- Homoantecessor: 800,000–200,000 years ago
- Homoheidelbergensis: 500,000–200,000 years ago
- HomoNeandertalensis: 300,000–28,000 years ago.
- Homofloresiensis: uncertain–18,000 years ago.
- Homosapiens: @200,000–present

Thus, we can see that homosapiens, being modern humans, appeared some 200,000 years ago and became the dominant species. All other hominid species were wiped out, and homosapiens became the most advanced of all living species on the planet.

The Out-of-Africa theory, which is strongly supported by anthropologists, argues that all modern humans evolved from a common female ancestor (University of Cambridge 2007) in Africa. This woman, known as "mitochondrial Eve," lived between 100,000 and 200,000 years ago in southern Africa. She was not the first ever female human but is considered to be the most recent woman from whom all living humans descend in an unbroken line purely through the female lines, until all lines converge on one woman. As a result, all humans today can trace their mitochondrial DNA back to her (New Scientist 2014).

Table 2.1 shows how humanity has grown in numbers over the millennia although the estimates prior to modern times are very speculative.

The table shows that the number of humans alive today is estimated at 8 billion, with future projections rising to 10 billion. However, that seems to be the ceiling, and demographers suggest that the human population will subsequently decline. Of course, there are many uncertainties surrounding this.

Natural Resources of the Earth

Natural resources refer to the things that exist freely in nature for human use and don't necessarily need the actions of humanity for their generation or production. The key aspect of natural resources is that they dictate the survival of humans and other life forms on Earth. These resources include:

- land,
- rocks,
- forests (vegetation),
- water (ocean, lakes, streams, seas, and rivers),
- fossil fuels,
- animals (fish, wildlife, and domesticated animals),
- minerals,
- sunlight,
- air.

Natural resources provide the basis of life on Earth, and it is from these resources that humans obtain and produce the components and materials found within our environment. These materials may be used as they naturally occur or may be transformed from other forms. It must be remembered that every artificial product is ultimately made from natural resources.

Natural resources can be categorized as renewable or non renewable:

- **Renewable natural resources:** These are the ones that are consistently available regardless of their use. They can be fairly easily recovered or replaced after utilization. Examples include vegetation, water, and air. Animals can also be categorized as renewable resources because they can be reared and bred to reproduce offspring to substitute the older animals. However, care is needed here. As much as these resources are renewable, it may take tens to hundreds of years to replace them. The renewable raw materials that come from living things, namely, animals and trees, are termed as organic renewable resources while those that come from nonliving things such as sun, water, and wind are termed as inorganic renewable resources.

Table 2.1 Human population on Earth

Period	Estimated human population
130,000 BC	200,000
10,000 BC	3 million
6500 BC	10 million
2000 BC	50 million
0 AD	200 million
1804	1000 million
1927	2000 million
1959	3000 million
1974	4000 million
1987	5000 million
1999	6000 million
2012	7000 million
2022	8000 million
2037	9000 million
2057	10000 million

Source: Various.

- **Nonrenewable natural resources:** These are the ones that cannot simply be substituted or recovered once they have been utilized or destroyed. Examples of such natural resources include fossil fuels and minerals. Minerals are categorized as non renewable because, although they take shape naturally through the rock cycle, their formation periods take thousands of years. Minerals are of particular relevance in relation to climate change for some of them are key ingredients for things like computers, mobile phones, batteries for electric vehicles, and so on.

There are many threats to the Earth's natural resources, and these are discussed in Chapter 5.

The Earth's Climate

First, let us consider what we mean by the term "climate." Climate is not the same thing as weather. The difference between them is an issue of time. Weather concerns the conditions of the atmosphere over a short period of time (i.e., today), and climate is how the atmosphere "behaves" over relatively long periods of time. Thus, for example, Spain might be regarded as having a generally warm and dry Mediterranean climate but occasionally the weather in Spain might be snow.

Climate is a word used to describe weather patterns using meteorological variables such as

- temperature,
- humidity,
- atmospheric pressure,
- wind speed and direction, and
- precipitation (rain, snow, etc.).

The climate is the overall state of the components of the climate system, including the atmosphere, the hydrosphere, the cryosphere, the lithosphere, and the biosphere, and the interactions between them.

The climate of the Earth is something which varies enormously across the planet and is affected by latitude, longitude, terrain altitude, nearby water sources and their currents. Just two examples illustrate this:

- **Temperature:** This can vary from –60°C in Antarctica to 50°C in the Middle East with an average global temperature of around 14°C.
- **Precipitation:** This averages 100 cm per annum across the globe but can vary from 10 cm in central Australia to 250 cm in the rainforests of Brazil or parts of central Africa.

In the early 1900s, a German climate scientist named Wladimir Koppen divided the world's climates into a series of categories. His categories were based on the temperature, the amount of precipitation, and the times of year when precipitation occurs. The categories were also influenced by a region's latitude being the imaginary lines used to measure our Earth from north to south from the equator.

Today, climate scientists split the Earth into five main types of climates. They are as follows:

- **Tropical:** In this hot and humid zone, the average temperatures are greater than 64°F (18°C) year-round, and there are more than 59 inches of precipitation each year.
- **Dry:** These climate zones are so dry because moisture is rapidly evaporated from the air, and there is very little precipitation.
- **Moderate (Temperate):** In this zone, there are typically warm and humid summers with thunderstorms and mild winters.
- **Continental:** These regions have warm to cool summers and very cold winters. In the winter, this zone can experience snowstorms, strong winds, and very cold temperatures, sometimes falling below –22°F (–30°C).
- **Polar:** In the polar climate zones, it is extremely cold. Even in summer, the temperatures here never go higher than 50°F (10°C).

The other dimension of climate is, of course, historical, and it is well known that the climate of the Earth has varied enormously over time. However, this is complicated and climate variations have always occurred in the short, medium, and longer term on a cyclical basis. Cyclical variations in the Earth's climate occur at multiple time scales, from years to decades, centuries, and millennia. Cycles at each scale are caused by a variety of physical mechanisms and climate, over any given period, is an expression of all of these mechanisms and cycles operating together.

Until fairly recently, over the 4.5 billion years of its life, the climate of the Earth was influenced solely by a number of physical factors with the impact of its human population being minimal. As already noted, during this huge period of time, the Earth's climate went through a number of cycles of change.

Thus, the Earth's climate has changed in many ways and in many cycles over periods of millions and billions of years. No doubt the Earth's climate will continue to change long into the future. However, as already noted, the key issue with climate change is, of course, not whether it is happening (it is) but what is causing it to happen. Are the changes in the future solely a matter of natural forces (as climate deniers maintain) or are they now a consequence of the actions of humanity.

Conclusion

The story of the development of the universe, the planet Earth, life on Earth, and human life is a fantastic and immensely rich story. We see how the Earth itself has changed in so many ways and through so many cycles over billions of years.

During the vast bulk of time since human life was created, humanity has posed little threat to the planet itself because it did not have the capability to do so. Today, humans and human activities constantly pose threats to the planet. The Centre for the Study of Existential Risk at Cambridge University is a research center intended to study possible existential-level threats posed to the Earth and humanity. Existential threat risks are defined as risks that threaten the destruction of humanity's long-term potential, which could either cause outright human extinction or irreversibly lock in a drastically inferior state of affairs. Such risks can be divided into two groups.

- **Nonanthropogenic:** risks outside human control such as asteroid impact, super volcanic explosion, a lethal gamma-ray burst, and so on.
- **Anthropogenic:** risks caused by humans and include nuclear war, human-induced pandemics, bioterrorism, and so on. To this list, must be added climate change.

We have already seen that we have no alternative home anywhere within traveling distance from the Earth, and we know of no other form of life in the universe. In spite of this, we are prepared to allow the degradation of our wonderful planet to a point where human existence may be very uncomfortable for many and almost impossible for most.

CHAPTER 3

Climate Change and Its Causes

Introduction

As already noted in Chapter 2, climate is not the same thing as weather. Weather concerns the conditions of the atmosphere over a short period of time (i.e., today), and climate is how the atmosphere "behaves" over relatively long periods of time. The term climate is used to describe weather patterns using meteorological variables such as temperature, humidity, atmospheric pressure, wind speed and direction, and precipitation (rain, snow, and so on).

This chapter discusses what is meant by climate change and how climate change is driven. It also covers some of the myths about climate change. It deals with the following issues:

- Earth's climate change history
- Global warming and climate change
- What is causing global warming?
- The role of greenhouse gas (GHG) emissions
- What are the future trends?
- Conclusion

Earth's Climate Change History

Unlike changes in weather, which are short term in nature, climate change is concerned with changes over relatively long periods of time. The first thing to note is that the climate of the Earth has never been constant and has changed many times, sometimes dramatically, throughout the Earth's 4–5 billion-year history. Long periods of stability, or equilibrium, were occasionally disrupted by periods of change that varied in length and intensity. Climatic shifts were destructive and, as we saw in Chapter 2, some even caused mass extinction events that wiped out high percentages of species. Despite these extinctions, life has always rebounded, allowing new species to dominate the landscape.

Some examples of extreme climatic conditions in the Earth's history include the following examples:

- **The Snowball Earth:** During one or more of Earth's icehouse climates, it is believed that the planet's surface became entirely or nearly entirely frozen. It is believed that this occurred sometime before 650 million years ago. There is no consensus as to what exactly caused these events, but one theory holds that a number of large volcanic eruptions sent sulfur gas particles into the atmosphere that reacted with solar radiation to produce a cooling effect.
- **Carboniferous Rainforest Collapse:** The Carboniferous period (305 million years ago) was known for its marshy forest communities inhabited by the ancestors of reptiles, mammals, and amphibians. It was also an "icehouse" period, in which permanent ice caps sat at the Earth's poles. But around 305 million years ago, levels of carbon dioxide increased substantially. This caused the planet to warm, dry out, and experience more intense seasonal fluctuations. Such a climate was intolerable for the Carboniferous rainforest plants, leading to a shift in the types of plant and animal communities and eventually the age of the dinosaurs. We should note that these GHG emissions came about naturally, unlike the GHG emissions which are generated by human activity.
- **Extinction events:** The most well-known example of extreme climate change is the extinction of the dinosaurs. Sixty-six million years ago, an asteroid collided with the Earth, sending a colossal cloud of ash and other debris into the atmosphere. This dense cloud blocked out the sun, creating an "impact winter" and halting the photosynthesis of plants and phytoplankton. The effects of the impact winter rippled throughout ecosystems, causing the extinction of the non-bird dinosaurs.

- **Thermal Maximum:** Over a period of about 100,000 years, the planet slowly warmed by between 5°C and 8°C. What caused the warming? Some scientists point to a volcanic eruption that prompted marine sediments to release methane, a powerful GHG, into the atmosphere. Oceans across the globe reached tropical temperatures, causing the extinction of a significant percentage of marine life.

These events probably have nothing in common other than being the outcomes of uncontrollable events that took place on Earth at that time, such as volcanic eruptions, increased solar radiation, asteroid impact, and so on. In all these examples, we see that geological phenomena and natural cycles can drastically alter the Earth's physical attributes. This includes the chemical composition of Earth's oceans and atmosphere, the wind and ocean currents, the ice caps, and other factors that contribute to Earth's climate. These shifts in climate—rainfall, temperature, sea level, and more—can, in turn, severely disrupt the ability of organisms and ecosystems to function.

Global Warming and Climate Change

Climate change is now probably one of the most commonly used phrases in human discourse today. To repeat, it is probably the biggest challenge facing humanity in its history. As we have discussed above, the climate of the Earth has changed throughout the whole history of the planet, and such changes have taken many forms and differing degrees of extremity. So why are we so concerned about climate change today.

The key issue driving climate change concerns rising temperatures on planet Earth, which constitutes what is now termed "global warming." To understand changes and variations in our climate, it is essential to know how the Earth's temperature changes from month to month, decade to decade and century to century.

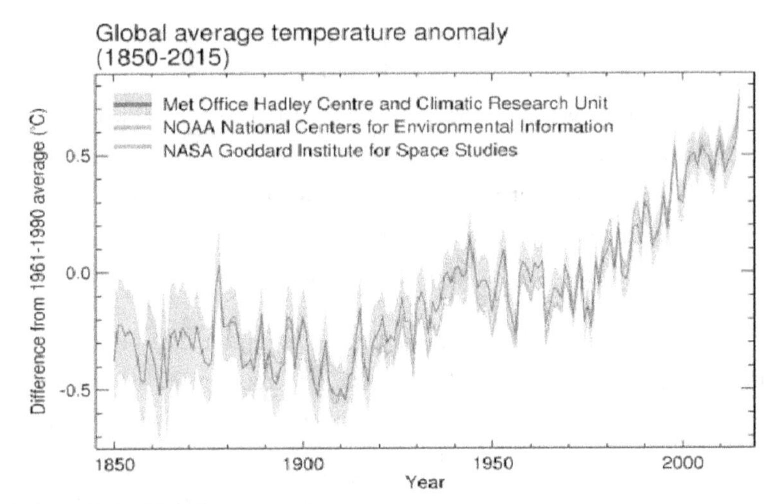

Figure 3.1 Global average temperatures

Recent History of Global Warming

Global-average surface temperature records provide this vital information. However, such data have only been available since 1850, as reasonably accurate thermometers did not exist prior to then, so we have only limited data. From these records, we can see how warm specific months, years, or decades were, and we can discern trends in our climate over longer periods of time.

There are three centers that calculate global-average temperatures each month.

- The UK Met Office
- The Goddard Institute for Space Studies (USA)
- National Climatic Data Centre (USA)

These organizations work independently and use different methods in the way they collect and process data to calculate the global-average temperature. Despite this, the results of each are similar from month to month and year to year, and there is definite agreement on surface temperature trends from decade to decade. Most importantly, they all agree global-average temperature has increased over the past century. This is illustrated in Figure 3.1.

The 170-year temperature record shows a lot of variation but can be divided into several major sections:

- A roughly constant average temperature in the late 19th century.
- A rise in the early 20th century
- A leveling off or slight decline in the mid-20th century
- A steep rise in the final decades of the 20th century, continuing into the 21st century

Clearly, we are talking here about the average surface temperature of our planet, and there will be big variations in surface temperatures in different parts of the Earth. However, temperatures are rising in many places on Earth, thus driving up the average temperature. The bottom line is that Earth's average temperature has risen about 0.65°C since the start of the 20th century, with nearly all regions of the globe experiencing a temperature increase. This may not seem like much of an increase, but we will see later why even such a small rise in the global average temperature is climatologically significant.

So, what causes this pattern of temperature variation, and how typical is it for our planet? A quick summary is as follows:

- Variations in the early part of the 150-year temperature record are largely explainable by natural variations within the climate system, volcanic eruptions, and variations in the Sun's energy output.
- Variations in recent decades are largely the result of human activities, predominantly but not exclusively the burning of fossil fuels.

However, there are other indicators of the Earth's temperature beyond the ones mentioned above. These include the following:

- Boreholes provide a record of temperatures in the recent past. Borehole temperatures smooth out the year-to-year fluctuations and provide a general trend consistent with surface temperatures.
- The heat content of the upper layers of the ocean shows an overall increase over the past few decades.

- Temperatures in the lower atmosphere (the *troposphere*), although harder to measure than surface temperatures, show an increase which is consistent with the rising surface temperatures.
- Temperatures in the upper atmosphere (the *stratosphere*) show a decrease which paradoxically is consistent with increasing surface temperatures (Real Climate 2006).
- Another indicator is the temperature of the oceans. Sea surface temperatures have been consistently higher during the past three decades than at any other time since reliable observations began in 1880.

Other Climatic Indicators

Aside from rising temperatures, there are a number of other indicators that point to global warming. A few, but nonexhaustive, examples include the following:

- ***Glaciers:*** Most mountain glaciers around the world are shrinking at accelerating rates. Glacier growth and shrinkage are determined by a balance between accumulating snow and the effects of melting.
- ***The Arctic:*** Sea ice is melting at accelerating rates. Melting sea ice results in a positive feedback, as darker ocean water absorbs more sunlight and results in further warming.
- ***Land-based ice caps:*** These are also melting. Greenland's ice melt is especially significant and is accelerating rapidly. The situation in Antarctica is less clear, with melting at the margins and some ice cap growth in the interior. There is also a concern that meltwater flowing under the ice may lubricate ice sheets, making them prone to slide into the sea, thus raising sea levels.
- ***Weather patterns:*** Tracking weather patterns over decades shows changes that would be expected from a changing climate. For example:
    - There are more extremely hot days and heat waves.

- Precipitation is increasing as warmer temperatures drive more evaporation, and precipitation is falling in intense but brief events.
- Hurricane intensity is increasing, especially in the North Atlantic, which seems seem to correlate with rising sea-surface temperature.
- Droughts are becoming longer and more extreme around the world.
- There is less snowpack in mountain ranges and polar areas, and the snow melts faster.

Longer Term Aspects of Global Warming

As we only have data from 1850, then clearly this is a limited period of time in the context of the Earth's history. How can we be sure that this isn't just a 150-year blip in a history of billions of years?

Fortunately, the lack of thermometers prior to 1850 does not mean that we have zero information about what had happened in previous centuries or millennia. Because there aren't enough accurate temperature measurements to calculate a global average temperature before the mid-19th century, scientists can use "proxies," meaning physical quantities that serve as indicators of temperature. Commonly used temperature proxies include the temperature:

- ***Annual tree rings***: The thickness and density vary with temperature and length of the growing season. Tree rings are most valuable in regions showing large winter–summer temperature variations.
- ***Coral reefs***: These form annual layers of calcium carbonate, whose analysis yields information about the temperature at the time of formation. Corals are most useful for tropical ocean temperatures.
- ***Lake sediments***: The thickness of these indicates the rate of snowmelt that feeds streams carrying sediment and, thus, provides a measure of springtime temperatures. Furthermore, the

pollen content of lake sediments tells us about plant species, giving a general indication of climatic conditions.

A number of independent studies have tried to reconstruct global temperature patterns back one or two millennia. From these, we see that the temperature reconstructions vary somewhat, but all show patterns of gradually declining temperature over most of the past millennium, followed by a sharp upturn through the 20th century and into the 21st century. These reconstructions suggest that Earth's current climate is the warmest in, at least, the past millennium.

Looking beyond this, ice cores from Greenland and Antarctica use oxygen and hydrogen isotope ratios to provide temperatures ranging back nearly a million years. These records show a cyclic pattern of brief (10,000 to 20,000 year) warm spells called interglacial periods, separated by longer cold spells (ice ages). For the past half million years, this cycle has repeated at roughly 100,000 year intervals. This pattern results from subtle changes in Earth's orbit and tilt, along with complex feedback effects in the climate system.

What Is Causing Global Warming?

To understand the answer to the above question, we first need a basic understanding of some principles of physics concerning matter and energy.

Firstly, with respect to matter, the Earth is essentially a closed system. With the exception of relatively small things like meteorites entering from space, probes going into space, and some leakage of the Earth's atmosphere into space, the materials that make up the planet and its atmosphere and oceans never disappear. Some cycle continually among different parts of the Earth system, and many change in chemical configuration. Take just one example, the element carbon. The carbon found on Earth might move, over a long period of time, from being carbon dioxide gas to being the composition of a diamond found in the ground to being part of protein molecules found in the human body. However, the total amount of carbon on Earth, at any point in time, is virtually unchanged.

This contrasts with energy. With respect to energy, the Earth isn't a closed system. Energy can enter the Earth and its atmosphere through the arrival of sunlight, but energy can also leave the Earth and go into space as infrared radiation. The incoming energy can be lesser or greater than the outgoing energy, with impacts on the Earth's climate. However, a stable climate requires a balance between incoming sunlight and outgoing infrared radiation. If the balance is disturbed because incoming energy remains constant while the outgoing energy increases, then the Earth would cool. However, in a situation where the incoming energy remained unchanged but the outgoing energy declines, the temperature of the Earth would rise, meaning global warming. This is the situation we find ourselves in because of what is referred to as "greenhouse gases" or GHG.

Greenhouse gases are gases in the Earth's atmosphere that produce what is termed the greenhouse effect. These gases surround and insulate the Earth like a blanket. They allow the sun to reach and warm the Earth's surface, then block the warmth from escaping back into space. As these gases continue to be emitted into the atmosphere, they form a thicker layer. And just like the blanket, the thicker it is, the more heat it stops. At this point, mention is often made of the planet Venus, which is a tremendously hot planet as a consequence of its own thick "blanket" that stops heat from escaping.

The Role of Greenhouse Gases (GHG)

Greenhouse gases occur naturally and are part of our atmosphere's makeup. For that reason, Earth is sometimes called the "Goldilocks" planet—its conditions are not too hot and not too cold, but just right to allow life (including us) to flourish. Part of what makes Earth so amenable is its natural greenhouse effect, which keeps the planet at a friendly 14–15 °C (57–59 °F) on average. But in the last century or so, human actions have been interfering with the planet's energy balance, mainly through the burning of fossil fuels that add carbon dioxide to the air. Human activities have continued to increase and have upset the balance of the natural system via emissions of several GHGs. The main GHGs are as follows:

- ***Carbon dioxide (CO_2):*** This is the biggest contributor to the problem. It occurs naturally from a number of sources. In the natural system, CO_2 can be absorbed by oceans and plants, taking it in during photosynthesis. The man-made increases in CO_2 have tipped the balance, and these natural intake sources cannot absorb these additional amounts.
- ***Methane (CH_4):*** This is the second largest culprit of the GHGs. In addition to the same industrial sources that emit CO_2, landfills and rice cultivation are additional sources of methane gas. While methane does not last as long in the atmosphere as CO_2, it is much more effective at trapping heat.
- ***Water vapor (H_2O):*** Water vapor is known to be Earth's most abundant GHG, but the extent of its contribution to global warming has been debated. Its role is more complicated to follow, involving the interplay between water vapor, carbon dioxide, and other atmosphere-warming gases. Basically, it is believed that water vapor can amplify the warming effect of other GHGs. The warming brought about by increased carbon dioxide allows more water vapor to enter the atmosphere. These warmer temperatures cause more water vapor to be absorbed into the air, and thus, warming and water absorption increase in a spiraling cycle.

Many material cycles are important to climate change, including those concerning carbon dioxide, water, and methane cycles, and the carbon, nitrogen, phosphorus, and sulfur cycles. But of these, water and carbon are the most important, in part because both water vapor and carbon dioxide are significant GHGs.

The Carbon Cycle

It was noted above that the Earth is a closed system with regard to matter. Hence, the amount of carbon on the planet never changes. However, the carbon is in a constant state of movement from place to place. It is stored in what are known as reservoirs. Carbon can be stored

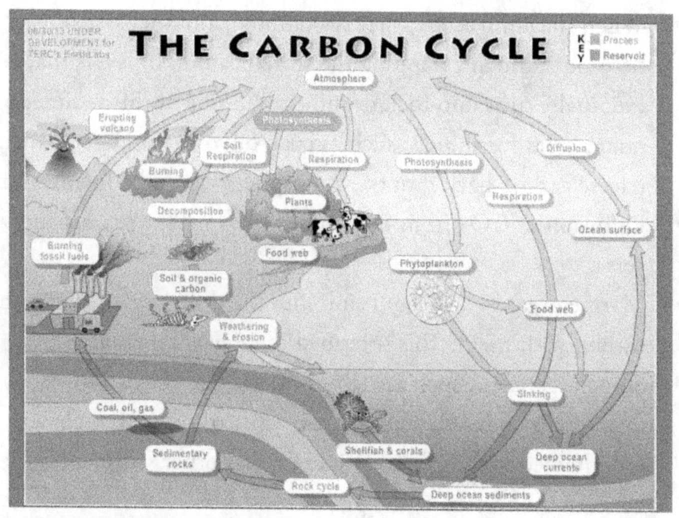

Figure 3.2 The carbon cycle

in a variety of reservoirs, but the amount of carbon in a specific reservoir can change over time as carbon moves from one reservoir to another. This is illustrated in Figure 3.2.

Thus, the *carbon cycle* operates on many timescales and involves interactions among atmosphere, land, oceans, biological systems, and on the longest timescales, geological processes. Figure 3.2 shows the carbon reservoirs and the processes by means of which carbon moves from one reservoir to another.

Some examples of what is taking place are as follows:

- Carbon in the atmosphere is captured by plants to make food during photosynthesis.
- Plants use the energy stored in carbohydrates for cellular activities and locomotion.
- Carbon can then be ingested and stored in animals that eat the plants.
- This process of *respiration* takes oxygen from the atmosphere and returns CO_2.
- Respiration, along with the decay of dead and waste matter, returns nearly all of the CO_2 that was removed in photosynthesis.

- When the animals die, they decompose, and their remains become sediment, trapping the stored carbon in layers that eventually turn into rock or minerals. Some of this sediment might form fossil fuels, such as coal, oil, or natural gas, which release carbon back into the atmosphere when the fuel is burned.
- Diffusion of CO_2 from the atmosphere into the oceans and vice versa.
- Marine organisms die and sink into the deep ocean, taking their carbon with them. This is termed the *biological pump*.
- Some organisms, such as clams or coral, use the carbon to form shells and skeletons. Most of the carbon on the planet is contained within rocks, minerals, and other sediment buried beneath the surface of the planet.

The atmosphere, living things, soils, and surface ocean waters all represent short-term carbon *reservoirs*. Cycling among these reservoirs occurs mostly on relatively short timescales. In particular, a typical carbon dioxide molecule remains in the atmosphere only about 5 years.

The problem is the carbon held in the deep reservoirs under the ground, which have been untouched for millions of years. For the past few centuries, humans have been removing oil, coal, and gas from the ground and burning it to provide energy. That process combines carbon with oxygen to produce carbon dioxide, which we release into the atmosphere through our chimneys. About half of that CO_2 accumulates in the atmosphere, resulting in atmospheric carbon dioxide rising nearly 40 percent since the start of the industrial era to levels the planet hasn't seen in at least a million years. Furthermore, we have released additional carbon through deforestation and agricultural practices, and land-use changes have altered the reflection of sunlight. All these processes affect Earth's climate, and recent climate change cannot be explained without including the human factor.

The Water Cycle

This involves the evaporation of surface water and its subsequent precipitation onto land and oceans. Solar energy ultimately drives the water cycle, with the rate of evaporation strongly dependent on temperature. The amount of water in the atmosphere is tiny compared to the amount stored in the oceans and cryosphere (frozen water). Although it varies with time and location, water vapor typically constitutes a few percent of the lower atmosphere. Atmospheric water vapor responds quickly to changing temperature and other conditions. From the long-term perspective of the climate, therefore, atmospheric water vapor can be considered to adjust almost instantaneously to changing surface conditions. Atmospheric water vapor contains the (latent) energy that was required to evaporate it, and therefore, the water cycle also contributes to the flow of energy from the surface to the atmosphere.

Methane

Two key characteristics determine the impact of different GHGs on the climate:

- The length of time they remain in the atmosphere
- Their ability to absorb energy

While carbon dioxide is more abundant and longer lived, methane (CH_4) is far more effective at trapping heat. Over the first two decades after its release, methane is more than 80 times more potent than carbon dioxide in terms of warming the climate system. Estimates of methane emissions are subject to a high degree of uncertainty, but the concentration of methane in the atmosphere is currently estimated to be around two-and-a-half times greater than its pre-industrial levels. The increase has accelerated in recent year. About 40 percent of these emissions are from natural sources, with the remaining 60 percent originating from human activity, known as anthropogenic emissions.

The largest anthropogenic source of methane is agriculture, responsible for around one-quarter of emissions, closely followed by the energy sector, which includes emissions from coal, oil, natural gas, and biofuels. Overall, methane is estimated to be responsible for around 30% of the rise in global temperatures since the industrial revolution.

What Are Future Trends in Global Warming and Impacts?

This is the difficult, controversial, and scary bit. We have seen the historic trend of rising surface temperatures on Earth over the last 20 years or so, but where is this leading us and what will be the impacts? (WEF 2021b).

Scientists and policy makers have long agreed that global warming beyond 2°C above the pre-industrial average would pose large and escalating risks to human life as we know it on Earth, and governments have used that number as an organizing principle. The Intergovernmental Panel on Climate Change (IPCC), founded in 1988, has published many detailed and rigorous reports, but it was not the first body to note the importance of limiting global warming to 2°C. In the 1970s, William Nordhaus, an economist at Yale, suggested in several papers that if global warming were to exceed 2°C on average, it would push global conditions past any point that any human civilization had experienced. At the time, Nordhaus's idea was a simple suggestion of how a rise in temperature could cause extreme conditions, based on the historical record of past average temperatures, but it gained new importance a decade later.

In 1988, amid mounting evidence that the Earth was warming, James Hansen, a NASA scientist, testified before Congress and became one of the first scientists to publicly link GHG emissions from humans to this warming trend. Hansen warned that if the world did not reduce emissions, it could result in catastrophic climate change, causing sea-level rise, extreme weather, and damage to ecosystems and human settlements across the globe. After Hansen's testimony, other groups of scientists started to study what might constitute "catastrophic climate change," and many papers used 1°C or 2°C as reference points to model what might happen to the Earth at different levels of warming.

Under the Paris Agreement at the COP in 2015, countries agreed to make plans to limit their emissions of GHGs. This agreement clearly defines 2°C as the upper limit for global warming but also lists 1.5°C as a more desirable goal because it reduces the risk of the worst outcomes of climate change in most of the world. With a 2°C increase, the risks of extreme heat waves, droughts, water stress, and extreme weather

Table 3.1 Projected temperature increases above the pre-industrial average

	Near term 2021–2040		Mid term 2041–2060		Long term 2081–2100	
Scenario	Best estimate °C	Very likely range °C	Best estimate °C	Very likely range °C	Best estimate °C	Very likely range °C
Very low emissions	1.5	1.2–1.7	1.6	1.2–2.0	1.4	1.0–1.8
Low emissions	1.5	1.2–1.8	1.7	1.3–2.2	1.8	1.3– 2.4
Intermediate emissions	1.5	1.2–1.8	2.0	1.6–2.5	2.7	2.1–3.5
High emissions	1.5	1.2–1.8	2.1	1.7–2.6	3.6	2.8–4.6
Very high emissions	1.6	1.3–1.9	2.4	1.9–3.0	4.4	3.3–5.7

Source: IPCC, page 14. www.ipcc.ch/report/ar6/wg1/downloads/report/IPCC_AR6_WGI_SPM_final.pdf

would be far greater for a larger portion of the Earth than with a 1.5°C increase. The IPCC now uses 1.5°C as a target in its reports rather than 2°C. The key message is that while both 1.5°C and 2.0°C are used as common frameworks for defining emission goals today, the current slow pace of international climate action has put the world in jeopardy of exceeding one or both targets.

The sixth assessment report of the IPCC, produced in 2022, provided a set of projections on how the Earth's temperature might change for the rest of this century using a variety of different scenarios. These were based on five shared socio-economic pathways (SSPs), which are scenarios of projected socio-economic global changes up to 2100. They are used to derive GHG emission scenarios with different climate policies. The projected temperature implications of these pathways are shown in Table 3.1, and they will be described further in Chapter 9.

Now, at this point, a word of warning is needed. This is pretty broad-based stuff, and the figures are "projections" not "predictions" and should be treated with caution. Forecasting temperature changes is difficult 5 years ahead, let alone 80 years ahead, given the complexity of the phenomena being dealt with and the number of variables involved. With more data and a better understanding of climate change, climate scientists will probably be able to refine these projections as time goes on. However, they are the best that can be done at this point in time, and they do paint a picture of what might happen if we ignore them at our peril. Another point is that these projections represent an average of global temperatures across the planet. In practice, of course, there will be considerable variations between different parts of the Earth and between seasons.

The following comments should also be made about these figures:

- The near-term projections show broad similarity for all five scenarios. This is because much of the damage regarding GHG emissions has already taken place and cannot be reversed in the short term. Indeed, there now seems to be an expectation that a 1.5°C rise will have taken place before 2030.
- Moving into the mid-term, both the best estimates and very likely ranges for projected temperatures start to diverge considerably.
- In the longer term, the projections show continual increases in temperature, which become alarming at the higher levels of emissions.

We now need to consider the potential impacts on our planet of temperature rises of the orders of magnitude shown above. Again, I must emphasize that much of this is speculative and lacks certainty, but there is much to consider and be concerned about. Let us first consider the issue of whether we should be aiming to constrain temperature rises to 1.5°C or 2.0°C. The first question, is does it matter? After all, it is only a difference of half a degree. Well, it does matter, and the following data from the Climate Council (2018) show the difference between 1.5°C and 2.0°C, illustrating this clearly.

- Global population exposed to severe heat at least once in every 5 years—a rise from 14 percent to 37 percent.
- Amount of Artic permafrost that will thaw—a rise from 4.8 million km to 6.6 million km.

- Reduction in yield from maize harvests in the tropics—3 percent reduction increasing to a 7 percent reduction.
- Decline in marine fishing: a reduction 1.5 million tones rising to a reduction of 3 million tones.

Similarly, evidence from the World Economic Forum (2021) shows that a 2°C rise will have the following impacts compared to 1.5°C rise:

- A doubling in the rate that plant species will be lost and a tripling of the rate that insect species will be lost
- A large increase in incidences of extreme heat;
- A 30 percent increase in the decline of coral reefs
- A tenfold increase in the existence of ice-free summers in the Artic sea

Now I have already emphasized the uncertainty surrounding these projections and I might also be accused of being selective in my examples. However, I would argue that other data examples will give the same picture and the key point is that the difference between 1.5°C and 2.0°C is not a minor issue but one which has huge impacts on the environment. This is illustrated in Table 3.2

If we look at larger increases in temperature, the situation becomes even more alarming as shown in Table 3.3.

Table 3.2 Impact of half a degree temperature increase

	1.5°C rise	**2.0°C rise**
% Global population exposed to severe heat at least once every 5 years	14%	37%
Amount of Artic permafrost that will thaw	4.8 million km	6.6 million km
Reduction in yield of from maize harvests in the tropics	−3%	−7%
% of plant species lost	−8%	−16%
% Decline in marine fishing	1.5 million tones	3.0 million tones

Source: Climate Council 2018.

Table 3.3 Impact of larger temperature increases

Temperature increase	Impacts
+ 2°C	• Water availability in some vulnerable regions (Southern Africa and Mediterranean) could drop by 20%–30% • Crop yields in Africa drop by 5%–10% • 40–60 million more people are exposed to malaria in Africa • Up to 10 million more people are affected by coastal flooding. Arctic species, including the polar bear and caribou, run a high risk of extinction • The Greenland ice sheet could begin an irreversible melt
+ 3°C	• In southern Europe, serious drought happens and occurs every 10 years • 1 to 4 billion more people suffer water shortages • Up to 5 billion gain waters but they could suffer increased floods Another 150 to 500 million people are at risk of hunger (if carbon fertilization is weak) • 1 to 3 million more people die from malnutrition. • Increased risk of abrupt changes in monsoons • A higher risk that the West Antarctic ice sheet, and the Atlantic thermohaline circulation, will collapse
+ 4°C	• Water availability in Southern Africa and Mediterranean could drop by 30%–50% • African agricultural yields drop by 15%–35% • Up to 80 million more Africans are exposed to malaria Another 7 to 300 million people are affected by coastal flooding
+ 5°C	• Some of the large Himalayan glaciers may disappear, affecting one quarter of the Chinese population and millions in India • Ocean acidity continues to rise; marine ecosystems are seriously disrupted • Sea level rise threatens small islands, low-lying coastal areas such as Florida, (Florida) and major world cities such as New York, London, and Tokyo

Sources: Various.

Conclusion

So where does all this leave us.

In the early part of the chapter, I explained that there is a lot of evidence and a strong consensus among climate scientists about the following three phenomena:

- The Earth is getting warmer.
- Much of this warming can be attributable to human activity and is not part of a natural climatic cycle.
- There will be severe consequences for the planet and its inhabitants associated with this level of warming.

However, while there is a lot of certainty that the above phenomena will, to some extent, take place, there is a lot of uncertainty about the magnitudes involved. This is because, as noted previously, we are trying to make projection decades ahead about a set of phenomena, which are not fully understood.

The problem is that climate change skeptics and deniers (which are discussed in Chapter 4) jump on this uncertainty and use it as an argument for doing nothing until better projections are available. There are two responses to this which need to be emphasized (and will be repeated in Chapter 4). These are as follows:

- Even the most optimistic view of the increase in global warming and its impacts will still lead to a lot of disruption and discomfort across the world. This will be much worse in poorer countries than rich countries.
- If we wait for more accurate projections to be available, it will be too late to take the remedial actions required to stabilize the climate.

Hence, urgent and radical action is required **now**.

CHAPTER 4

Are We Certain That Climate Change Is Real?

Introduction

Although there is widespread belief and acceptance, across the globe, for the existence of global warming leading to climate change, there is also a strong lobby in many countries that argues that global warming and climate change are myths and something we shouldn't worry about.

Scientific knowledge is vitally important in resolving this climate change debate, but we need to clear about the distinction between

1. robust scientific findings,
2. tentative scientific findings,
3. distorted scientific findings,
4. opinions,
5. advocacy,
6. gossip and hearsay, and
7. media stories.

I would suggest that in considering claims about the reality of climate change, Categories 3–7 should be treated with extreme caution, especially those that come from many dubious sources. Category 2 should also be treated with slightly less caution, pending further research, but the most important evidence is that in Category 1, namely, robust scientific findings. This is easier said than done since few people would be able to access, read, and understand original scientific papers on climate change, and so they are reliant on others to interpret the findings for them. Unfortunately, these interpretive sources are not always honest in doing this.

Research-based evidence suggests a very strong consensus exists among climate scientists, from a multitude of political, social, organizational, and national backgrounds, about the extent and cause of climate change. Some researchers have concluded that around **97** percent of climate scientists agree with this consensus (Cook et al. 2013), and while there might be some skepticism about the accuracy of this figure, it is clear that the vast majority of climate scientists are in broad agreement about the current position and likely trends regarding global warning and climate change.

However, a word of warning is necessary here. The climate of the Earth is an immensely complex system with numerous inter-relationships between variables. Hence, the reality is that the workings of our climate are not fully understood, but this understanding will improve over time. In the meantime, climate scientists are trying to make projections of future global warming trends for years and decades into the future and, at present, they are working with incomplete models of climate change behavior. Thus, at the present time, it is unreasonable to expect absolute precision in their projections, and it is important to look at the broad picture, direction, and pace of travel.

This chapter will cover the following:

- The nature of climate change skepticism and denial
- Categories of climate change denial
- Characteristics of climate change deniers
- Conclusion and warning

The Nature of Climate Change Skepticism and Denial

In spite of the scientific evidence available, there are many people around the globe who deny or doubt the existence of climate change and/or the role of human actions in its creation. They prefer to believe that climate change is natural and not caused by human actions and that the current global warming is just part of the normal cycle of events that the Earth has gone through in its 4 billion-year history.

There are different perspectives among climate change skeptics or deniers, but these can be grouped twofold as follows:

1. People who are skeptical about global warming (or who deny its very existence) believe that no significant climate warming is taking place, while also claiming that the warming trend measured by weather stations is an artifact due to urbanization around those stations (the so-called urban heat island effect).
2. Attribution skeptics or deniers (who accept the global warming trend but see only natural causes for this) doubt that human activities are responsible for the observed trends. A few of them even deny that the rise in the atmospheric CO_2 content is anthropogenic (i.e., caused by human activities) or that additional CO_2 does not lead to discernible warming and that there must be other natural causes for warming.

More specifically, climate change deniers make a large number of false or unfounded claims that climate change is not the problem it is perceived to be and does not require the drastic actions being proposed. Shown below are some examples of the false claims made.

- Some climate change denial groups say that because CO_2 is only a trace gas in the atmosphere (roughly 400 ppm, or 0.04 percent, 4 parts per 10,000), it can only have a minor effect on the climate. This is very misleading, and scientists have known for over a century that even this small proportion of CO_2 has a significant warming effect, and doubling the proportion leads to a large temperature increase. The scientific consensus, as summarized by the IPCC fourth assessment report, the U.S. Geological Survey, and other reports, is that human activity is the leading cause of climate change. The burning of fossil fuels accounts for around 30 billion tons of CO_2 each year, which is 130 times the amount produced by volcanoes.
- Some groups allege that water vapor is a more significant GHG and is left out of many climate models. While water vapor is a GHG, the very short atmospheric lifetime of water vapor (about 10 days) compared to that of CO_2 (hundreds of years) means that CO_2 is the primary driver of increasing temperatures. Also,

as noted earlier, water vapor acts as a feedback, not a forcing, mechanism in relation to climate change.

- Some often point to natural variability, such as sunspots and cosmic rays, to explain the warming trend. According to them, there is natural variability that will abate over time, and human influences have little to do with it. The reality is that these factors are already taken into account when developing climate models, and the robust scientific consensus is that they cannot explain the observed warming trend.
- While not exactly denying climate change, some commentators suggest that we should wait for better technologies to come on stream before addressing climate change, when they will be more available, affordable, and effective. In my opinion, it is really "pie in the sky" to bank on future technologies that might not come to fruition and/or will be less effective than anticipated. As a young child, I remember reading the promise that controlled nuclear fusion would provide humanity with an endless source of energy in the not-too-distant future. Sixty years later, we are still some way off from fulfillment of that promise.

Climate change denial is a very serious issue, and I spend some time looking at it because failure or success in undermining the arguments and data in relation to global warming and climate change will have profound implications for actions to cope with climate change and the future of humanity.

Categories of Climate Change Denial

There are very many statements and claims made of different kinds alleging that global warming and climate change aren't taking place. Some of these claims are made in good faith, but others can be seen as devious, misleading, and based on vested interests. However, it is useful to try and categorize these denial claims into a small number of categories. John Cook (2015) has provided a framework for categorizing climate change denial approaches, which are discussed below.

Fake Experts

As already mentioned, 97 percent of climate scientists agree that humans are causing global warming. This is based on a number of studies, including surveys of scientists, analysis of public statements about climate change, and analysis of peer-reviewed scientific papers. Then how might one cast doubt on the overwhelming scientific consensus? One technique is the use of "fake experts."

One example of this concerns online petitions such as the Global Warming Petition Project, which featured more than 31,000 scientists claiming humans aren't disrupting our climate (Global Warming Petition Project n.d.). How can there be a 97 percent consensus when 31,000 scientists disagree? The answer is that the vast bulk of the petition's signatories aren't climate scientists at all. Only 0.1 percent of the scientists who signed the petition are climate scientists, and only 0.5 percent is in the related fields of atmospheric science and/or meteorology. So why aren't these 0.1 percent who are climate scientists being challenged about their views? Because they are not identified.

If you were suffering from chest pains, would you consult a dermatologist—no, you would consult an expert cardiologist. It is the same with climate change. What is the point of having a petition signed by other types of scientists, including computer scientists, mechanical engineers, and medical scientists, who would have only a limited knowledge of climate science? This is an example of fake experts in bulk.

Another example concerns a politician. At a May 2018 meeting of the United States House Committee on Science, Space, and Technology, a House Representative claimed that sea level rise is caused not by melting glaciers but rather by coastal erosion and silt that flows from rivers into the ocean (CNN 2018). A subsequent scientific calculation showed that it would take the equivalent of the top five inches of land from the entire surface area of the United States to cause the oceans to rise the 3.3 mm a year currently seen. And those five inches would have to fall into the oceans every year to explain the current rate at which they are rising. This was clearly a ludicrous and unsubstantiated claim pronounced by a high-profile non-scientist.

Logical and/ or Factual Fallacies

As already noted, there is a very strong consensus among climate scientists about the reality of global warming and climate change, not least because there are many lines of evidence pointing in the same direction. Aside from the evidence of rising surface temperatures, there are also observations of heat escaping out to space, in the structure of the atmosphere, and even in the changing seasons.

Another denialist technique used to counter the weight of evidence is the use of logical and/or factual fallacies. If we consider possible statements, it is possible to judge these statements from two directions—logic and fact. The possible results of this are illustrated in Figure 4.1.

Thus, we see that by combining the logical and factual aspects of a statement, there is only one combination which can be deemed correct and that is a statement which is both correct in logic and fact. Climate deniers often tend to ignore one or both of these aspects when talking about climate change.

One common fallacious logical argument is that current climate change must be natural because climate has changed naturally in the past. This myth commits the logical fallacy of jumping to non-evidenced conclusions. This is like saying that smoking cigarettes cannot be the cause of lung cancers since people died from lung cancer before smoking was invented. The premise does not lead to the conclusion. In a similar vein, some Australian politicians tried to deny the impact of global warming on Australian climate (The Guardian 2013). One stated "Australia has had fires and floods since the beginning of time. We've had much bigger floods and fires than the ones we've recently experienced. You can hardly say they were the result of anthropogenic global warming." This false argument distracts from the fact that the risk of fires in Australia has been rising across many Australian locations since the 1970s. Fire danger days are happening not just in summer but also in spring and autumn.

Turning to factually incorrect statements, there are many examples that can be quoted. One is that plants need carbon dioxide (CO_2) to live, so we shouldn't worry about rising levels of CO_2. Plants and forests do need CO_2 for growth, and in doing so, they remove and store away huge amounts of CO_2 from the atmosphere each year. But the problem is,

		Logic: Correct	Logic: Incorrect
Fact	Correct	Logically and factually correct	Logically incorrect but factually correct
	Incorrect	Logically incorrect but factually correct	Logically and factually incorrect

Figure 4.1 Logical and factual correctness

there's only so much CO_2 they can absorb, and this amount is getting less as more and more forests are cut down across the world, largely to produce our food. Another factual error is that polar bear numbers are increasing in spite of climate change. This is a vast oversimplification of a complex issue. Climate change is the biggest threat faced by polar bears. The Arctic is warming roughly twice as fast as the rest of the world, causing sea ice to melt earlier and form later each year. This makes it more difficult for female polar bears to get onto land in late autumn to build their dens, and more difficult for them to get out onto the sea ice in spring to feed their cubs. Their main source of prey, seals, is also affected by climate change, as they depend on sea ice to raise their young. However, a recent survey showed that the status of polar bear populations was as follows:

- Four populations are in decline.
- Two populations are increasing.
- Five populations are stable.
- Eight populations are data-deficient (information missing or outdated)

So, while polar bear numbers are increasing in 2 out of 19 populations, the overall picture is bleak (WWF 2021). By 2040, scientists predict that only a fringe of ice will remain in Northeast Canada and Northern Greenland when all other large areas of summer ice are gone. This "Last Ice Area" is likely to become important for polar bears and other life that depends on ice. A projection of sea ice in the archipelago shows that much of the region is facing significant ice loss in the coming decades with potentially serious consequences for polar bears. Global polar bear numbers are projected to decline by 30 percent by 2050—a very different picture than that stated at the beginning of this section.

Cherry Picking

Signs of global warming have been observed all over our planet, including the following examples:

- Ice sheets in Greenland and Antarctica are losing hundreds of billions of tons of ice every year.
- Global sea levels are rising.
- Thousands of species are migrating toward cooler regions in response to warming.
- The ocean is building up four nuclear bombs worth of heat every second.

One way to avoid this overwhelming body of evidence for global warming and climate change is through the technique of "cherry-picking." To try and prove a point, climate change deniers pick on and utilize some small and unrepresentative sample of data and draw conclusions from that which ignore the bigger picture. For example, one persistent view is that global warming slowed in recent years. While this is factually true up to a point, it is extremely misleading. It is true that global surface temperature rises declined slightly in 2021, but they continued their upward march in both 2022 and 2023 (NOAA 2023). Some other key points to remember about this example are as follows:

- It just focuses on one aspect of our climate system, namely, the surface temperature records, and ignores others measures of temperature which were discussed in the previous chapter (e.g., ocean temperatures) and which are also rising.
- It just considers a relatively short period of time, not the longer term trends.
- It ignores the many warming indicators telling us that our planet continues to build up heat.

Impossible Expectations

While many lines of evidence inform our understanding of climate change, another source of understanding is climate models. These are computer simulations built from the fundamental laws of physics, and they have made many accurate predictions since the 1970s. Climate models have successfully predicted the loss of Arctic Sea ice, sea level rise, and the geographic pattern of global warming. However, another technique used to cast doubt on climate models is the tactic of impossible expectations.

Unrealistic expectations occur when some people argue that climate models are unreliable if they don't make perfect short-term predictions. A number of unpredictable influences, such as ocean and solar cycles, have short-term influences on climate, but over the long term, these effects average out, which is why climate models do so well at long-term predictions rather than short-term.

Hence, while climate change models do provide useful longer term projections, the uncertainties involved mean it is impossible to provide the level of precision looked for by climate change deniers. In time, with a better understanding, short-term predictions might also improve.

Conspiracy Theories

This is perhaps the most difficult aspect of climate change denial to deal with. A conspiracy theory can be defined as an explanation for an event or situation that invokes a conspiracy by certain groups when other explanations are more probable. Conspiracy theories abound in our society and include the following:

- The human moon landings were fake.
- President John Kennedy was not killed by a lone gunman.
- Jeffrey Epstein was killed in prison and did not commit suicide.
- President Barack Obama was not born in the United States.
- Many countries are controlled by a "deep state" of powerful interests which subvert the official government.
- The Covid epidemic was deliberately engineered by the Chinese government, and so on.

Now we can be reasonably certain that virtually all governments and government officials would take the view that none of the above conspiracies are true, and indeed many of them do look absurd. However, the problem is that, in spite of the denials, many people still have doubts about whether some of these conspiracies have a ring of truth about them. The best example here concerns the assignation of President John Kennedy. The Warren Commission was set up in 1963 to investigate the assassination, and a key finding was that Lee Harvey Oswald acted alone and unaided. Fifty years after the assassination, a poll indicated that 61 percent of people believed that Oswald did not act alone, and another poll taken showed that 62 percent of people thought that there was an official cover-up to keep the public from knowing the full truth. These issues become even more fascinating when one considers who might have gained from the assassination event taking place.

There are also many conspiracy theories about climate change, including the following:

- A plot to test a US secret weapon. A Russian political scientist suggested that talk of climate change might be intended to facilitate tests of US weapons that could cause "droughts, erase crops, and induce various anomalous phenomena in certain countries."
- A statement by a Chinese author that global warming is a plot hatched by the West to deprive the developing world of its legitimate share of the world's wealth.
- A globalist plot intended to bring about a world government.
- An excuse to promote the further development of nuclear power and thus make Britain's energy industry independent of the trade unions.
- A plot to replace the free and spontaneous evolution of mankind by a sort of central, global planning of the whole world.
- A plain and simple scam which involves scientists peddling scare stories to chase funding.

One looks at some of these conspiracies and wonders how people can believe them, but I suggest that the problem is multi-faceted, including a declining trust in governments and other authorities, which are often seen to lie about various matters. Also, there is the situation that some of the nonclimate conspiracies listed earlier are seen as being potentially true in spite of repeated denials by authorities. This sort of thing can spill over into climate change conspiracies.

The link between conspiratorial thinking and science denial has serious and practical consequences. Conspiracy theorists are often immune to scientific evidence since any evidence conflicting with their beliefs is considered part of a conspiracy. The implication is that the most effective approach is not changing the mind of the unchangeable. It is suggested that a more fruitful approach is communicating the realities of climate change to the large, undecided majority who are open to scientific evidence. Also important is the need to consider and point out who gains from a particular conspiracy. As I consider later, some of those promoting climate denial will potentially gain from climate mitigation actions being stopped. Whether this is adequate remains to be seen.

Characteristics of Climate Change Deniers

In a court of law, jurors and judges will listen to what witnesses say, and in drawing conclusions from the evidence given, will focus on what the witness has said but also the background of the witness and whether their evidence can be regarded as reliable or unreliable. In deciding reliability, they will consider the background of the witness and whether they have a track record of lying, but also whether they might have some vested interest in what happens to the person on trial. For example, it the witness was seen as likely to gain if their evidence led to a conviction of the defendant, then one might have some doubt about the veracity and reliability of their evidence.

We need to consider the above point when considering the backgrounds and motives of those loudly denying the existence of climate change. There are a number of issues to be aware of.

Nationalists

Nationalist politics abounds today in virtually every corner of the globe. However, it has been suggested that climate change may conflict with nationalist political views because it is "unsolvable" at the national level and requires collective action between nations, which is something nationalists are wary of. Consequently, nationalism sometimes tends to reject the science of climate change, which is a serious problem. The problem is that nationalism has no solution to climate change, and if you want to be a nationalist in the 21st century, you have to deny the climate problem. It is still possible to be patriotic, and there is still room in the world for having special loyalties and obligations toward your own people and toward your own country. But in order to confront climate change, you need additional loyalties and commitments to a level beyond the nation. This means international collaboration.

Political Ideologies

For many years, climate change was seen as a purely environmental issue which had implications for the environment and the planet on which we live. Consequently, at that time, it was not a significant ideological issue. The adoption of the 1997 Kyoto protocol on climate change altered all that. What was previously a scientific issue became one which impacted powerful political and economic interests in many countries.

The basic ideological divide is probably between the political right, which is skeptical of climate change, and the political left, which sees it as a major problem. This disagreement is shown in Figure 4.2 for selected countries.

In the United States, there is a huge divide between Democrats and Republicans on the issue of global warming and climate change. In January 2019, a survey found that 67 percent of Democrats (and Democratic-leaning independents) said dealing with global climate change should be a top priority, compared to just 21 percent of Republicans and Republican-leaning independents—a 46-point gap. Also, while 74 percent of Democrats said protecting the environment should be a top priority, only 31 percent of Republicans said the same. Divisions on the importance of these two issues were larger than on other key issues such

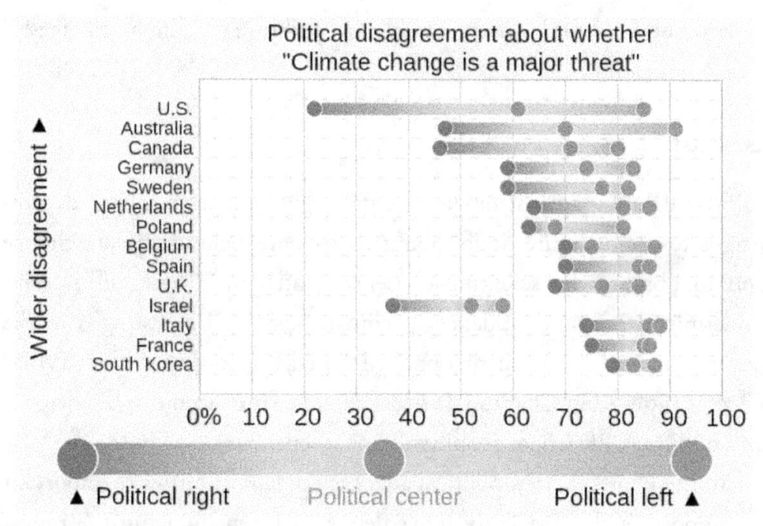

Figure 4.2 Political disagreement over climate change

Source: Poushter et al. (2022).

as strengthening the military and dealing with the issue of immigration. Similar divisions can be seen in Canadian and Australian political groups.

In the UK, the situation is much more nuanced than this. The Conservative party has many MPs who strongly support measures to deal with issues of climate change. However, a section of Conservatives in Parliament, as well as many party members, would not rank tackling the climate crisis as a major priority. While few, if any, UK parliamentarians will publicly deny climate change is happening, there are those prepared to chip away at the idea by suggesting that it is not as bad as being claimed.

The reasons for this divide between the political right and left are complex and multi faceted but probably are linked to a number of factors such as:

- The impact of climate change actions on economic growth, jobs, and prosperity.
- Views on the merits of "big" government involvement in climate change actions.
- The need for cooperation on internationally determined climate change measures as opposed to nationally determined measures.

- The need for large-scale public expenditure on climate changes measures.

Lobbyists

Efforts to lobby against environmental regulation have included campaigns to manufacture doubt about the science behind climate change and to obscure the scientific consensus and data. These efforts have undermined public confidence in climate science. Interestingly, in the quotations at the start of this book, I noted the comment from Prince Charles (now King Charles III) that climate change seems to be the only area of human life where people will not accept scientific advice.

In the United States, political advocacy organizations were important in supporting the Tea Party movement and in encouraging the movement to focus on opposing climate change actions. Other conservative organizations were also significant participants in these lobbying attempts, seeking to halt or eliminate environmental regulations.

This approach to downplay the significance of climate change was copied from tobacco lobbyists who, in order to delay the introduction of regulation, tried to undermine the scientific evidence linking tobacco to lung cancer. Lobbyists also attempted to discredit the scientific research by creating doubt and manipulating debate. They worked to discredit the scientists involved, to dispute their findings, and create and maintain an apparent controversy by promoting claims that contradicted scientific research.

Business Corporations

It is often thought, by many people, that most business corporations are largely hostile toward actions taken to deal with the problems of climate change. This is a simplistic view. Business corporations may have a difficult balancing act when considering their attitude toward global warming and climate change. There are three aspects to consider.

- **Adaptive response:** Some businesses recognize that the impact of climate change (and the measures introduced to counter climate change) will have a serious impact on their business but not necessarily an existential impact. This could involve such things as rising energy costs, shortages of certain materials or higher prices for those materials, unacceptable production methods, and so on. In these circumstances, the business can adapt what it does to deal with these climate change impacts and enable it to continue in business. A good example of this might be automobile manufacturers who had to recognize that, because of global warming, the internal combustion engine was doomed, and to survive, they had to shift to the manufacture of electric or hydrogen vehicles.

- **Customer perceptions:** The declared public attitude of a company toward climate change may impact the way it is viewed by current and potential customers. This is particularly the case with those customers who are supportive of the need for action against climate change. Hence, companies may wish to take actions, and publicize those actions, to show that they are serious about reducing their own contribution to global warming.
- **Strong resistance:** Some actions to mitigate climate change may have major (and perhaps existential) impacts on company business prospects. This leads them to take a strong stance against actions to mitigate climate change, which will impact, seriously, on their business. Included in this would be actions opposing climate change measures and campaigning against such actions, including support for climate change denial. Examples here would be actions taken by oil and coal companies who see serious (and perhaps existential) long-term damage to their existing business activities. For example, several large corporations within the fossil fuel industry have attempted to mislead the public about the trustworthiness of climate science. Between 1989 and 2002, the Global Climate Coalition, a group of mainly United States businesses, used aggressive lobbying and public relations tactics to oppose action to reduce GHG emissions and fight the Kyoto Protocol. The coalition was financed by large corporations and trade groups from the oil, coal, and auto industries. The New York Times reported that

"even as the coalition worked to sway opinion toward its own scientific and technical experts were advising that the science backing the role of GHGs in global warming could not be refuted." Recently, we have seen attempts by oil-producing countries at the 2023 COP summit to undermine climate change measures.

The issue of climate change and business is further discussed in Chapter 10 of the book.

Conclusion and Warning

The thrust of this chapter is that while there is a very strong scientific consensus about the reality of global warming and climate change, there is still a significant body of opinion that is skeptical about, or rejects entirely, that this is the case. I have outlined some of the ways in which climate deniers support their claims (e.g., cherry picking of data), but perhaps the most difficult and insidious aspect is that of conspiracy theories. As I have noted, all of the scientific evidence in the world will not refute conspiracy theories since the scientific community is regarded as part of the conspiracy. I have to ask myself why climate scientists are devoted to scientific research would want to produce (as climate change deniers allege) false research evidence in order to win research grants to produce yet more false evidence.

While much of the activity of climate deniers is undertaken honestly in the (mistaken) belief that climate change is not as bad as it is made out to be, there is a large number of climate deniers who, insidiously, try to destroy the credibility of scientific evidence to further their own personal agenda.

Whether climate change denial can be neutralized is unknown factor, but as it stands, it is a major barrier to getting people understand the seriousness of climate change and the need for action.

CHAPTER 5

Climate Change Isn't the Only Problem

Introduction

As already noted, the issue of climate change is the single biggest challenge that has faced humanity throughout its history. However, there are three other phenomena that are taking place, simultaneously, and that are strongly inter-related to climate change and each other. These are as follows:

- The loss of biodiversity in nature
- The degradation and destruction of natural habitats of life
- Loss of nonrenewable natural resources

The three phenomena are interrelated in several directions and in different ways. In this chapter, I will explore these phenomena and the links between them more thoroughly.

I discuss these issues as follows:

- Loss of biodiversity
- Habitat destruction
- Loss of nonrenewable natural resources
- Inter-relationships
- Conclusion

Loss of Biodiversity

Basically, the term biodiversity concerns the variety and variability of life on Earth. It considers the numbers and variations of plants, animals, and microorganisms residing in a variety of vastly different ecosystems such as oceans, forests, coral reefs, grasslands, tundra, deserts, rainforests, and so on. Biodiversity includes not only species that are rare, threatened, or endangered but also every living thing from humans to organisms, which we know little about, such as microbes, fungi, and invertebrates.

According to the United Nations Environment Programme (UNEP), variation in biodiversity is typically measured at three different levels.

- The ecosystem level.
- The species level.
- The genetic level

However, biodiversity is not evenly distributed on planet Earth, which shows huge variations. It is richest in the tropics, and tropical forest ecosystems contain about 90 percent of the world's species but cover less than 10 percent of Earth's surface. Marine biodiversity tends to be highest in areas with high sea surface temperatures, including the mid-latitudinal band in all oceans and the areas along coasts in the Western Pacific.

However, biodiversity is under threat in many parts of the world and I discuss the following issues in this chapter:

- Why is biodiversity important?
- The scale of biodiversity loss on Earth
- What are the main causes of loss of biodiversity?
- The impact of biodiversity loss

Why Is Biodiversity Important

Biodiversity is important to most aspects of our lives. Healthy ecosystems provide us with many essentials we take for granted. For example:

- Plants convert energy from the sun making it available to other life forms.
- Bacteria and other living organisms break down organic matter into nutrients providing plants with healthy soil to grow in.
- Pollinators are essential in plant reproduction, guaranteeing our food production.

What many do not realize is that biodiversity directly affects humans as well. Loss of biodiversity disrupts the essential mechanisms needed for food production, health maintenance, and climate regulation. In short, biodiversity provides us with clean air, fresh water, good quality soil, and crop pollination. It helps us fight climate change and adapt to it as well as reducing the impact of natural hazards. Since living organisms interact in dynamic ecosystems, the disappearance of one species can have a far-reaching impact on the food chain. It is impossible to know exactly what the consequences of mass extinctions of species would be for humans, but we do know that, for now, the diversity of nature allows us to thrive.

The importance of healthy ecosystems and rich biodiversity on Earth can be illustrated by the following examples, which are not exhaustive.

- ***Soil formation and protection:*** A greater variety of plants in an ecosystem helps soil and makes it rich in nutrients. In turn, plants store these nutrients, which are consumed by animals and finally given back to the environment when they die.
- ***Carbon sequestration:*** Carbon sequestration is the process of capturing and storing carbon dioxide from the atmosphere. It reduces atmospheric carbon dioxide, and its ultimate goal is to reduce climate change. Vegetation and soil in ecosystems like forests, peatlands, grasslands, seabed's, wetlands, and kelp act as carbon sinks, removing carbon dioxide from the atmosphere.
- ***Food resources:*** A greater variety of plants and poultry animals result in more food resources in a nation and globally. This is vital in a world that struggles to feed its population. Our food system and agriculture are strongly linked to biodiversity. Millions of species work together to supply us with a variety of grains, vegetables, fruits, and animal products. Food production relies on many "services" that biodiversity provides. This includes pollination, maintenance of soil fertility, resistance to pests and diseases, climate maintenance, and water filtration.
- ***Ecosystem efficiency:*** Each species in an ecosystem has a specific role to play. Most of these are interdependent on each other for

their survival. Thus, a reduction in biodiversity in an ecosystem will impact on its overall efficiency.

- ***Disease resistance:*** Genetically diverse populations of plants or animals have better chances of surviving a catastrophe like a pandemic. Diverse populations carry genetic codes that make certain members of their group less vulnerable. When those carrying these genetic codes reproduce, disease resistance is passed along, and the species' survival is strengthened.
- ***Storm, flooding, and coastal erosion:*** Coastal sea levels are rising, and the World Economic Forum (WEF 2021 (a)) says that as many as 410 million people could be affected by the end of the century. While 59 percent of sea level rise is expected to be in tropical Asia, countries such as China, France, Senegal, Nigeria, and the United States are also at risk. Restoration and protection of coastal ecosystems such as salt marshes and mudflats will be an important aspect of flood prevention for low-lying coastal communities. Coupled with greater root biomass, these ecosystems can provide better resistance to soil erosion. Ecosystems like coral reefs, seagrass, and soft-bottom ecosystems work as buffers against waves or storms, protecting coastal communities that are prone to typhoons.
- ***Medications:*** The medicinal property of plants is important for the pharmaceutical industry which will, therefore, be affected by loss of diversity.
- ***Freshwater resources:*** Biodiversity protects freshwater resources and keeps them clean.
- ***Overall well-being:*** Walking on grass, going to the mountains, or swimming in the sea, being in contact with nature has a host of well-being benefits for humans. Biodiversity offers locations for this activity. Places with greenery and flowing rivers, mountains, and beaches offer great recreation facilities for humans. Exposure to green and blue spaces outdoors improves our working memory, attention control, and cognitive flexibility. Research suggests that contact with nature is associated with

increased positive social interactions, happiness, and a sense of meaning in life, as well as decreases in mental stress.
- ***Unpredictable events:*** Biodiversity can help with such events. Greater species diversity ensures improved natural sustainability for all life forms and healthy ecosystems that can better withstand and recover from a variety of disasters.

Overall, it can be seen that biodiversity is vitally important and brings a huge range of benefits to almost everyone. Conversely, loss of biodiversity will have harmful effects in a variety of ways and in some cases (e.g., pandemics) the results could be catastrophic.

The Scale of Biodiversity Loss on Earth

We have to recognize that throughout the Earth's history, millions of species of life have been created and millions have become extinct and have disappeared. Nothing stays the same. It has been estimated that of the four billion species of life which have evolved on Earth during its history, 99 percent have already become extinct (National Geographic 2019). This will continue. Dinosaurs became extinct as a consequence of an asteroid impact but other well-known and more extinctions include the Dodo, the Black Rhino, and the Passenger Pigeon. Also, extinction rates are not constant, and in the Earth's history, there have been, at least, five mass extinction events including the one which killed off the dinosaurs.

Such extinctions can be natural or human created. Natural extinction occurs when a species declines in numbers gradually but steadily at the end of its evolutionary period on Earth. The length of the period depends largely on the success of the species as a whole and its ability to adjust to changes in climate and vegetation and the appearance of predators or (in the case of predators themselves) the disappearance of prey. It is worth remembering that a species dying a natural evolutionary death is nearly always replaced by new forms.

The list of known recent extinctions is still a small fraction of all species that have lived on the planet. Extinctions have always occurred but the rate at which they are happening now far exceeds the rates at which species have naturally gone extinct over the course of the fossil record. The historical spread of humanity over the planet has been associated with waves of extinctions in other species. Key threats to date have been over-hunting and harvesting of species by people, habitat conversion, and degradation, and the introduction of invasive species caused by human migration, settlement, trade, agriculture, and resource extraction. These threats have been accelerating in recent history alongside rapid growth in human populations and increasing growth in per capita consumption of resources. In addition, in the past few decades, climate change has become an increasingly important threat. Estimates suggest that extinction rates in the recent past have been running tens to hundreds of times faster than in pre-human times and that the pace is accelerating.

What Are the Main Causes of Loss of Biodiversity?

The following are a list of the main causes of biodiversity loss, many of which are easily recognizable.

- **Destruction of habitat:** The natural habitat of animals is destroyed by man for the purpose of settlement, agriculture, mining, industries, construction of highways, and so on. As a result of this, the species must either adapt to the changes in the environment or move to other places. If not, they become target to predation, starvation, disease, and eventually die.
- **Hunting:** Hunting of wild animals is done for the commercial utilization of their products. These include hides and skin, fur, meat, tusk, cosmetics, perfumes, pharmaceuticals, and decoration purposes. In recent years, 95 percent of the black rhino population in Africa has been exterminated by poachers for their horn. In addition to this, over one-third of Africa's elephants have been killed in the last decade to collect 3000 tons of ivory. Although the formulation of international and domestic laws and regulations has reduced hunting to some extent, a large amount but poaching still continues to be a threat to biodiver-

sity. Another problem, of course, is that many of the countries where these species reside are poor countries and do not have the resources needed for protection.

- **Habitat fragmentation:** A separation of parts of a habitat into spatially segregated fragments that are too limited to maintain their different species for the future. The landmass is broken into smaller units, which eventually lead to the extinction of species.
- **Collection for zoo and research:** Animals and plants are collected for zoos and biological laboratories. This is done for research in science and medicine. Primates such as monkeys and chimpanzees are sacrificed for research because of their anatomical, genetic, and physiological similarities to human beings.
- **Introduction of exotic species:** A species which is not a natural inhabitant of the locality is deliberately or accidentally introduced into the ecosystem. Due to the introduction of exotic species, native species have to compete for food and space.
- **Pollution:** Pollution can make survival difficult for a species as it alters its natural habitat. Examples are water pollution and toxic wastes that disturb the food chain. In addition, materials like insecticides, pesticides, sulfur, and nitrogen oxides, and acid rain also adversely affect the plant and animal species.
- **Control of pests and predators:** Generally, nontarget species that are a component of balanced ecosystem may also get killed in the predator and pest control measures.
- **Natural calamities:** Floods, draught, forest fires, earthquakes, and other natural calamities sometimes take a heavy toll of plant and animal life. These trap a large number of animals while frittering away soil nutrients.

The Impact of Biodiversity Loss

As already noted, we are losing biodiversity on planet Earth at an alarming rate, and this is occurring within species, ecosystems, large geographic area, and the Earth as a whole. This will have huge implications for humanity and the planet.

We can value biodiversity for many reasons, some utilitarian and some intrinsic. This means we can value biodiversity both for what it provides to humans and for the value it has in its own right. In utilitarian terms, loss of biodiversity can have the following impacts:

- **Human health:** Healthy ecosystem products and services (such as availability of fresh water, food, and fuel sources) are requisite for good human health and productive livelihoods. Biodiversity loss can have significant direct human health impacts if ecosystem services are no longer adequate to meet social needs.
- **Agricultural production:** Loss of biodiversity has a negative impact on agriculture. The more species there are, the greater diversity and resilience to changes in farming practices that we have for our food sources. With less plant variety, crops can be wiped out by disease or insects faster than they would if many plants were available as options. This can make it hard for farmers to maintain a sustainable farm with any level of profit because so much is lost when one crop fails. Loss of biodiversity creates an environment where only certain types of foods will grow well, which then limits what humans eat. The introduction or elimination of certain plants and animals can have dramatic effects on what crops are available to farmers, how much they must invest in pesticide use, which pests will be able to thrive without competition from other species, and so on.
- **Climate change:** As a result of human interference, there are many plants and animals around us with less genetic variation than they would have had in nature alone. Loss of biodiversity due to human activities like deforestation and ocean acidification is not only a problem for the animals themselves but also has an impact on climate change. We are losing too many species that help offset our release of carbon dioxide into the atmosphere, exacerbating global warming in turn, while their presence would have helped stabilize temperatures by absorbing some CO_2.
- **Economic output:** An overlooked impact of the loss of biodiversity is a possible decrease in economic productivity.

> Ecosystems and their biodiversity support the global economy and human well-being. Biodiversity contributes directly and indirectly to many constituents of human well-being, including security, basic material for a good life, health, good social relations, and freedom of choice and action. A loss of biodiversity has a significant impact on us, socially, and it may have a direct and indirect impact on the economy. Climate impacts like severe drought or excessive flooding have been shown to result in decreased agricultural production, which can lead to higher food prices for consumers worldwide.

Habitat Destruction

All forms of life exist in some habitat whether that be on land, in rivers, in oceans, in animals, or even some extreme locations such as the upper atmosphere, deep ocean trenches, and inside volcanoes. However, many of these habitats are facing degradation and even destruction, which will impact on the forms of life that exist in them. Thus, habitat destruction can be defined as the elimination or alteration of the conditions necessary for animals and plants to survive which impacts on not only individual species but also the health of the global ecosystem. Habitat destruction is one of the biggest threats facing plants and animal species throughout the world. The loss of habitat has far-reaching impacts on the planet's ability to sustain life.

Habitat destruction can occur in both land and water environments.

Degradation of Land

Land degradation involves the reduction or loss of the biological or economic activity resulting from land uses or a process (or combination of processes) arising from human activities and habitation patterns. Examples are as follows:

- Soil erosion caused by wind and/or water
- Deterioration of the physical, chemical, biological, or economic properties of soil and long-term loss of natural vegetation

Land degradation is caused by multiple forces, including extreme weather conditions, particularly drought. It is also caused by human activities that pollute or degrade the quality of soils and land utility. It negatively affects food production, livelihoods, and the production and provision of other ecosystem goods and services. Desertification is a form of land degradation by which fertile land becomes desert.

Coastal and Marine Degradation

The degradation of coastal and marine areas in a country is the result of a number of pressures and factors. Indiscriminate habitat conversion and other destructive practices including massive land reclamation or dump-and-fill procedures are common, involving the overloading of waste, including heavy sedimentation from mining and indiscriminate dumping of garbage.

Loss of Nonrenewable Natural Resources

In Chapter 2, I listed the natural resources of the Earth and explained that while some of these resources are renewable, some are not and, once used, have gone forever.

There are many threats to the Earth's natural resources, the key ones of which are:

- **Overpopulation:** We have already seen the enormous growth in the human population over the last hundred years or more. Basically, more humans mean a growth in the demand for natural resources, which may become unsustainable.
- **Pollution:** Environmental pollution is a leading cause of natural resource degradation and depletion. It is mainly caused by industries that produce and use chemicals and plastics in their operations. These chemicals slip into the soil and water systems and alter the composition of the resources.
- **Agricultural practices:** The growth in the numbers of humans, plus overconsumption by people in the developed world, has led

to enormous increase in food requirements on Earth. To meet those needs, a range of damaging agricultural practices have been used such as the use of harsh chemicals and deforestation to increase the amount of arable land. This results in the degradation of natural resources.

- **Modern lifestyles:** To maintain modern lifestyles, more and more natural resources are needed to meet the many demands humans have. This includes fossil fuels to generate energy and raw materials to produce consumer goods, and so on. Subsequently, these natural resources have been overused resulting in their depletion.
- **Frivolous activities:** Some natural resources that are nonrenewable are being frittered away by frivolous activities. Take, for example, helium gas. Helium gas is a critical component in many fields such as scientific research, medical technology, hi-tech manufacturing, space exploration, and national defense. However, we see helium gas being used to inflate balloons which to amuse people at parties, and so on. The point is that helium is a nonrenewable resource. Once it is gone, it is gone for good. There is no factory on the Earth manufacturing helium to replace what is being used.

Inter-Relationships of These Phenomena

Climate change, diversity loss, and habitat destruction must not be seen as separate phenomena. There are strong relationships between climate change and the other three issues discussed in this chapter, namely, those of loss of biodiversity, habitat destruction, and resource depletion. Moreover, there is a cause-and-effect relationship between these factors, which can work in opposite directions.

This is discussed in the sections below using suitable examples.

Climate Change and Biodiversity Loss

Climate change is disturbing natural habitats and species in ways that are still only becoming clear. There are signs that rising temperatures are affecting biodiversity, while changing rainfall patterns, extreme weather events, and ocean acidification are putting pressure on species already threatened by other human activities. One major impact of climate change on biodiversity is the increase in the intensity and frequency of fires, storms, or periods of drought. In Australia at the end of 2019 and start of 2020, 97,000 km^2 of forest and surrounding habitats were destroyed by intense fires that are now known to have been made worse by climate change. Rising global temperatures also have the potential to alter ecosystems over longer periods by changing what can grow and live within them. There is already evidence to suggest that reductions in water vapor in the atmosphere since the 1990s has resulted in 59 percent of vegetated areas showing pronounced browning and reduced growth rates worldwide.

On the other hand, we also know that natural habitats and biodiversity play an important role in regulating climate and can help to absorb and store carbon. Mangroves are significant sinks for carbon, and the Amazon is one of the most biologically diverse places on the planet and is an enormous store of carbon. Safeguarding these natural carbon sinks from further damage is an important part of limiting climate change. Loss of biodiversity can also accelerate climate change processes, such as the capacity of degraded ecosystems to assimilate and store CO_2.

Climate Change and Habitat Destruction

Coral reefs are important ocean habitats and offer a vivid case of the risks of climate change. Reefs provide a large fraction of Earth's biodiversity, and they have been called "the rain forests of the seas." Scientists estimate that 25 percent of all marine species live in and around coral reefs, making them one of the most diverse habitats in the world. Climate change is the greatest global threat to coral reef ecosystems. The Earth's atmosphere and ocean are warming, and as temperatures rise, mass coral bleaching events and infectious disease outbreaks are becoming more frequent. Coral calcification is the rate at which reef-building corals lay down their skeletons and is a measure of coral growth, which is important for healthy reef ecosystems. Carbon dioxide absorbed into the ocean dissolves in the warmer water and increases the acidity of the water. This has already begun to reduce calcification rates in reef-building and reef-associated organisms. Climate change will also affect coral reef ecosystems, through sea level rise, changes to the frequency and intensity of tropical storms, and altered ocean circulation patterns. When combined, all of these impacts dramatically alter ecosystem function, as well as the goods and services coral reef ecosystems provide to people around the globe.

The world's forests store enormous quantities of carbon. This is captured from CO_2 in the air and turned into the structure of the tree. Microorganisms in the soil draw even more carbon into the forest floor and left undisturbed; this carbon is safely locked away. But when the trees are cut down (and often burned), the carbon is released back into the atmosphere. As well as storing carbon, forests also create their own weather systems. The trees and other vegetation circulate water vapor to create clouds and rain. Taking the trees away also takes away those weather systems, which can lead to desertification Remember; 6000 years ago the Sahara was a tropical jungle.

Thus, we can see that forests play many important ecological roles in inhibiting global warming and climate change. Forests cover 31 percent of our planet with the Amazon rainforest alone being home to hundreds of thousands of species of plants and animals. Yet, despite everything forests do for the planet, they are being cut down at an alarming rate. Every year thousands of square miles of forest are lost to deforestation, equal to an astounding 11 football fields of forest per minute. Since 1990, the world has lost 420 million hectares or about a billion acres of forest, mainly in Africa and South America. About 17 percent of the Amazonian rainforest has been destroyed over the past 50 years, and losses recently have been on the rise. Figure 5.1 illustrates the scale of the problem.

Habitat Destruction and Biodiversity Loss

Natural habitats are being destroyed by humans for settlement, agriculture, mining, industries, highway construction, dam building, and so on. As a consequence, the species in those habitats must either adapt to the changes in the environment, move elsewhere or succumb to predation,

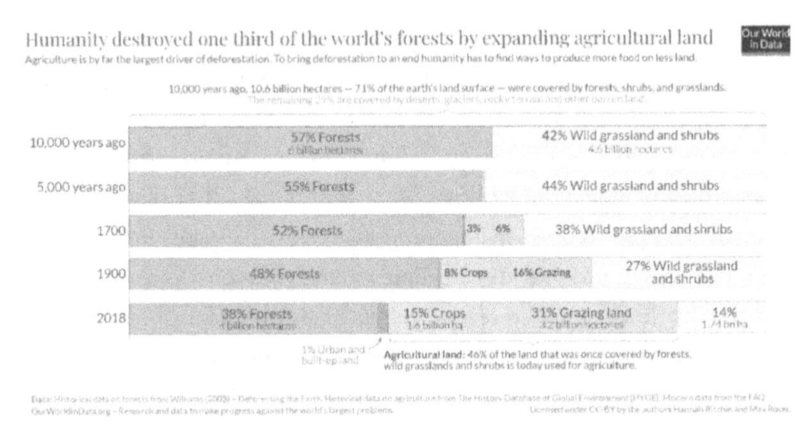

Figure 5.1 Destruction of the world's forests

Source: Amazon conservation

starvation or disease and eventually die. Several rare butterfly species are facing extinction due to habitat destruction in the Western Ghats, a mountain range in India. The cultivation of tea, coffee, rubber, and palm oil in the area has been propelling the slow destruction of the lush natural beauty of the region, once densely covered by forests. Human activities pose the greatest threat to the environment of the Western Ghats that negatively affect the region's rich and highly vulnerable biodiversity. The growing numbers of threatened and endangered species are clear evidence of the escalating situation. For example, of the 370 butterfly species available in the Ghats, around 70 are at the brink of extinction.

There are also situations where loss of biodiversity from some external cause can lead to a negative impact on the habitat of a species. An example of this can be seen in the apple and pear orchards of Sichuan, China, where the loss of insect pollinator diversity threatened the existence of the habitat. The absence of pollinators has been caused by the overuse of pesticides that are toxic to insect pollinators. As a consequence, laborers in this region are now armed with a paintbrush and a pot of pollen and must pollinate each individual flower by hand, which may not be an economically viable solution due to the rising cost of labor and declining fruit yields.

Natural Resources and Climate Change

The need to take action over climate change can also pose a threat to natural resources. For example, the mass introduction of electric vehicles and aircraft will produce huge demands for elements such as cobalt and lithium of which there is a limited supply in the Earth. The mining of these minerals also produces huge land degradation and pollution.

Also, projects are now being considered to mine for these minerals from the seabed of the Pacific Ocean. The impacts of these actions are incalculable in terms of their impact on habitats and biodiversity in the oceans.

Conclusion

I have already discussed, in earlier chapters, the issues of global warming leading to climate change on Earth. I noted that the human activities that have driven and are still driving climate change are enormous. In this chapter, we have seen how the related phenomena of biodiversity loss, habitat destruction, and resource depletion are closely inter-related with climate change but also added to, and reinforce, these pressures. Hence in considering the issues of global warming and climate change, we also need to have great regard for biodiversity loss and habitat destruction both in their own individual right and in relation to the other phenomena.

CHAPTER 6

Stopping Climate Change

What Is Supposed to Be Happening?

Introduction

Climate change has a long history, and scientists have been interested in variations in the Earth's temperature for several centuries. However, in the late 1950s, geochemist Charles Keeling started to record CO_2 levels from an observatory situated in the Pacific Ocean. Data from this observatory revealed what would become known as the "Keeling Curve." The upward, saw tooth-shaped curve showed a steady rise in CO_2 levels, along with short, jagged up-and-down levels of the gas produced by repeated wintering and greening of the Northern Hemisphere. In the 1960s, early computer modeling began to predict possible outcomes of the rise in CO_2 levels made evident by the Keeling Curve. These models consistently showed that a doubling of CO_2 could produce a warming of 2°C within the next century. However, the models were preliminary, and a century seemed a very long time away, so little was done.

In the early 1970s, a different kind of climate worry took hold: global cooling. As more people became concerned about the pollutants being emitted into the atmosphere, some scientists theorized that the pollution could block sunlight and cool the Earth. In fact, the Earth did cool somewhat between for a short period due to a postwar boom in aerosol pollutants, which reflected sunlight away from the planet. The idea that sunlight-blocking pollutants could chill Earth caught on in the media, as in a 1974 Time magazine article titled "Another Ice Age?" However, this brief cooling period ended, and temperatures resumed their upward climb.

The early 1980s would mark a sharp increase in global temperatures. Many experts point to 1988 as a critical turning point when watershed events placed global warming in the spotlight. The summer of 1988 was the hottest on record (although many since then have been hotter) and also saw widespread drought and wildfires within the United States. Scientists started sounding the alarm about climate change and the media and the public began to pay attention. One year later, in 1989, the Intergovernmental Panel on Climate Change (IPCC) was established under the United Nations to provide a scientific view of climate change and its political and economic impacts.

As the issue of climate change is now an important issue on the agenda of politicians and the general public in all countries, this chapter provides a discussion of what is actually happening (or planned to happen) to deal with the problem involved.

- The pivotal importance of "Net Zero"
- Levels of action on climate change
- Types of climate change action
- History of the Conference of the Parties (COP)
- The scale of the problem
- Areas for action
- Conclusions

The Pivotal Importance of "Net Zero"

As conversation around climate change grows louder, we have had to become familiar with the language used to describe what it all means. We hear terms like net zero, carbon neutral, and climate neutral, making their way into our vocabulary. But do these terms refer to the same thing? You would be forgiven for being confused as they seem to be used interchangeably by a variety of people. The answer is yes and no. In broad terms, all three definitions are concerned with reducing the amount of Anthropocene (human induced) GHGs getting into and staying in the atmosphere, but the precise definitions do vary. This is concerning because where there is uncertainty about what a term means, people may end up claiming to have achieved something which isn't correct.

What Is Net Zero?

Simple definitions of three related terms are as follows:

- **Carbon neutral:** Carbon neutrality is achieved when carbon emissions are balanced by the same amount of carbon removals from the atmosphere. Such carbon removals do not include carbon offsetting through mitigation elsewhere.
- **Net zero:** In contrast to mere carbon neutrality, the definition of "net zero" also includes neutrality of other GHGs such as methane and nitrous oxide. As with carbon neutrality, to reach net zero, the GHGs emitted into the atmosphere must be equivalent to the GHGs being removed from the atmosphere.
- **Climate neutral:** This term is often used as substitute for "carbon neutral," but according to its definition by the IPCC is actually comprising a far wider scope. It should include all activities that have an effect on the global climate, including water systems and land use. Since almost all human action has an effect on climate, true climate neutrality is almost impossible to claim.

In this chapter, I will be focusing on net zero as it is the most commonly used term.

Net Zero and Climate Change

Net zero has become the defining lens through which governments, businesses, and individuals view the fight against climate change. Net zero refers to a situation where the GHGs going into the atmosphere are balanced by removal of GHGs out of the atmosphere. The term net zero is important because this should be the state at which global warming stops. Thus, net zero is the internationally agreed goal for mitigating global warming in the second half of the century.

The IPCC concluded that the need for net zero by 2050 should remain consistent with limiting growth in surface temperature to 1.5°C.

The "net" in net zero is important because it will be very difficult to reduce all emissions to zero on the timescale needed. As well as deep and widespread cuts in emissions, we will probably need to scale up removals. In order for net zero to be effective, it must be permanent. Permanence means that removed GHG do not return into the atmosphere over time, for example through the destruction of forests or improper carbon storage. Permanent or hard "net zero" refers to a balance between all GHG sinks and sources that is sustained over matching time scales. To reach net zero, emissions from homes, transport, agriculture, and industry will need to be cut. In other words, these sectors will have to reduce the amount of carbon they put into the atmosphere. But in some areas, like aviation, it may be too complex or expensive to cut emissions altogether and these 'residual' emissions will need to be removed from the atmosphere: either by changing how we use our land so it can absorb more CO_2, or by being extracted directly through technologies known as carbon capture, usage, and storage.

Levels of Action on Climate Change

Actions to deal with the problems of global warming and climate change (plus the associated phenomena of loss of biodiversity, habitat destruction and resource consumption) can be addressed at several different levels of which the following may be identified.

- **Global:** Global warming and climate change cannot be addressed, effectively, by any one single nation since the activities of one country may spread around the world and affect others. While some nations have a much bigger impact on climate change than others, it is important that all nations are seen to be playing a part in what is a global issue of such huge magnitude. Hence, the formation of the Conference of Parties (COP). The COP is an international climate meeting held each year by the United Nations. Those countries who joined are "party to," in legal terms, the international treaty called the U.N. Framework Convention on Climate Change (UNFCCC). Parties to the treaty have committed to take voluntary actions to prevent "dangerous anthropogenic (human-caused) interference with the climate system." Countries take turns hosting an annual meeting at which government representatives report on progress, set intermediate goals, make agreements to share scientific and technological advances of global benefit, and negotiate future pledges.

- **National:** We have already seen that the UNFCC and the COPs have the responsibility for getting country agreements on matters like pledges to take actions to counter climate change, to provide funds for certain purposes, and so on. However, it is the responsibility of individual countries to take the necessary actions to implement these changes. This could involve country governments in such things as legislative changes, policy changes, enhanced regulation, provision of public funds, encouragement to citizens to change behaviors, and so on.

  Now climate science is not an exact science, and there are things that are not fully understood. Also forecasting the impact of certain actions on global warming is also inexact. Nevertheless, based on what is the best scientific knowledge, we might be reasonably confident that *implementing* the various COP pledges should enable global average temperature rises to be constrained to 1.5°C or 2.0°C. However, there are potentially two problems which may occur.

  1. Some countries have not signed-up to certain COP pledges or are half-hearted in their acceptance. We will see this in the discussion of COP later in the chapter.
  2. Although signed-up to pledges, the countries involved do not implement, quickly or adequately enough, the necessary actions needed to meet the pledges.

As we will see later, the whole approach is rather loose. However, the Glasgow Climate Pact has created new systems to keep the pressure on national governments. For instance, there is a so-called "no-cheat charter" that standardizes the method countries must use when self-reporting their emissions. Historically, many states have taken liberties with their self-reporting of emissions to the point that an investigation found emissions were at least 16 per cent higher than national reports would indicate.

- **Multi-national**—This is not the same as global. Whereas the COP has overall responsibility for the direction of global policies to mitigate climate change and nation states have responsibilities to take actions, there are examples of where several countries can collaborate together. Perhaps the best example of this concerns the European Union which takes responsibility for climate change policy across the 27 member countries. Its European Green Deal aims to transform the EU into a modern, resource-efficient and competitive economy, ensuring:
    - No net emissions of GHGs by 2050.
    - Economic growth decoupled from resource use.
    - No person and no place left behind.

- **Organisational**—By organizations I mean companies, public sector bodies, voluntary organizations, etc. It is important for such organizations to consider what actions they might take to counter global warming (and associated phenomena). This could involve looking at a variety of things such as waste recycling, use of energy, use of transport, and so on. Such actions may be important to ensure the sustainability of the organization itself and the way it is perceived, by customers, service users, general public, and so on. Most organizations (especially the larger ones) will already have policies and strategies for countering global warming.
- **Individual**—Most individuals these days will have some knowledge of the issue of climate change and some understanding of what actions they might take at the individual level (e.g., less traveling, more recycled waste, etc). Some will argue that this is pointless since the required change can only take place at the national or global level, not the individual level. However, it is important not to decry the attitude of "doing your bit" since I firmly believe that solutions to climate change, ultimately rely on changes in individual behaviors.

- **Community**—Finally, actions can be taken at the community level. This involves voluntary groups being set up in communities in order to publicize the need for climate change actions and to promote improvements in the area involved. There are good examples of such groups but, perhaps, they are rather small in number.

Types of Climate Change Action

In broad terms, there are two types of action which can be undertaken in relation to climate change. These are mitigation and adaptation, and these are outlined below. It is important to understand the distinction between them and I can start doing this by means of a story shown below.

Imagine your boat has sprung a leak. To keep from sinking you must address the source of the problem. That means plugging the holes. But what about all the water already rushing in? To stay dry, you grab a bucket and start bailing. To stay afloat and prevent damage to your boat, you need to address both issues simultaneously

Mitigation

Mitigation efforts require measures to address the underlying problem of climate change by taking action to slow down or stop the rise in fossil fuel emissions. Some examples of mitigation measures that can be taken to avoid the increase of pollutant emission include the following:

- Greater energy efficiency
- Shift to of renewable energy
- Electrification of industrial processes
- Reduced and cleaner means of transport
- Less consumption of food especially meat

Adaptation

Adaptation involves actions needed to help people and governments withstand and minimize the ravages of climate change that are already here and will worsen as the years go by. In some sense, adaptation can be seen as a sort of failure in that having failed to stop global warming, we now take measures to deal with the problems that are occurring via adaptation measures that help reducing vulnerability to the consequences of climate change. Some examples of adaptation measures include the following:

- More secure facility locations and infrastructures
- Landscape restoration and reforestation
- Flood prevention measures
- Flexible and diverse cultivation to be prepared for natural catastrophes
- Preventive and precautionary measures (evacuation plans, health issues, and so on)

It will come as no surprise that adaptation measures will be extremely expensive and will put pressure on the public finances of high-income countries let along low-income countries.

History of the Conference of the Parties (COP)

The Conference of Parties (COP) is the top-level decision-making body of the United Nations Climate Change Framework Convention (UNFCCC). The UNFCCC was formed in 1994 to stabilize the GHG emissions and to protect the Earth from the threat of climate change. COP members have been meeting every year since the year 1995 with the latest COP being COP 28 held in 2023. Over this period, the profile of the COPs has increased, as has the level of urgency concerning its work and the level of seniority of those attending. Shown below is a very brief summary of the earlier COPs, with greater detail about more recent events.

COPs 1–12

Early COP meetings were rather limited in ambition and actions, with frequent disagreements. There was also a lack of engagement by some countries. One exception was the Kyoto Protocol agreed in COP 3.

COPs 13–25

The COPs started to add some details to the Kyoto agreement. However, a lot of discussion took place about a more ambitious climate agreement when the Kyoto protocol expired.

COP 26 Glasgow 2021

A new global agreement was reached that aimed to reduce the worst impacts of climate change. However, some leaders and campaigners said it did not go far enough. Also, most commitments made at COP will have to be self-policed, and only a few countries are making their pledges legally binding.

Although not legally binding, the agreements would set the global agenda on climate change for the next decade, and the main points are as follows:

- **Emissions:** It was agreed that countries will meet to pledge further cuts to emissions of CO_2. This is to try to keep temperature rises within 1.5°C, which scientists say is required to prevent a "climate catastrophe." Current pledges, if met, will only limit global warming to about 2.4°C.
- **Coal:** For the first time at a COP conference, there was an explicit plan to reduce the use of coal, which is responsible for 40 percent of annual CO_2 emissions. However, countries only agreed a weaker commitment to "phase down" rather than "phase out" coal after a late intervention by China and India.
- **Developing countries:** A pledge to significantly increase money to help poor countries cope with the effects of climate change

and make the switch to clean energy. While some observers say that the COP26 agreement represented the "start of a breakthrough," other countries felt not enough progress was made.

- **Fossil fuel subsidies:–** An agreement to phase-out subsidies that artificially lower the price of coal, oil, or natural gas. However, no firm dates were set.
- **US–China Cooperation:** On paper, this was a big breakthrough. The world's two biggest CO_2 emitters, the US and China, pledged to cooperate more over the next decade in areas including methane emissions and the switch to clean energy. China has previously been reluctant to tackle domestic coal emissions; so this was seen as recognizing the need for urgent action.
- **Trees:** Leaders from more than 100 countries—with about 85 percent of the world's forests—promised to stop deforestation by 2030. This is seen as vital, as trees absorb vast amounts of CO_2. Similar initiatives haven't stopped deforestation, and it is unclear how the pledge will be policed.
- **Methane:** Methane is estimated to be currently responsible for a third of human-generated warming. A scheme to cut 30 percent of methane emissions by 2030 was agreed by more than 100 countries but the big emitters China, Russia, and India didn't join.
- **Private finance:** A large amount of private finance was proposed in order to facilitate initiatives such as renewable energy and direct finance away from fossil fuel-burning industries. The initiative is an attempt to involve private companies in meeting net zero targets. However, there is some skepticism as to whether this will actually happen, and it could be little more than a PR exercise.

COP 27 Sharm El Sheik 2022

In his stark opening address, the UN Secretary-General António Guterres said that more needs to be done to drastically reduce emissions now. He added that the world still needs a giant leap on climate ambition, and the red line we must not cross is the line that takes our planet over the 1.5°C temperature limit. "We can and must win this battle for our lives," he concluded. Unfortunately, there is a belief that the battle to keep below 1.5°C is already lost.

Some key points from the conference are as follows:

- **Loss and damage:** After three decades, there was an agreement to establish a fund to help countries seeking financial assistance to deal with the impact on the physical and social infrastructure of countries devastated by extreme weather. However, there was no agreement yet on how the finance should be provided and where it should come from.
- **Temperature targets:** The 2015 Paris agreement contained two temperature goals—to keep the rise "well below 2.0°C above pre-industrial levels and, pursuing efforts to keep the increase to 1.5°C". Science since then has shown clearly that 2° is not safe, so at COP26 in Glasgow, countries agreed to focus on a 1.5°C limit. As their commitments on cutting GHG emissions were too weak to stay within the 1.5°C limit, they also agreed to return each year to strengthen them, a process known as the ratchet. Some countries tried to renege on the 1.5°C goal and to abolish the ratchet. They failed, but a resolution to cause emissions to peak by 2025 was taken out, to the dismay of many.
- **Low emissions energy:** This could be interpreted as being a boost to many things, from wind and solar farms to nuclear reactors, and coal-fired power stations fitted with carbon capture and storage. It could even be interpreted to mean gas, which has lower emissions than coal but is still a major fossil fuel. This all seemed rather vague.
- **Fossil fuels:** After almost three decades of conferences on climate change, at COP 26 in Glasgow, a commitment to phase down the use of coal was agreed. Some countries led by India wanted to go further and include a commitment to phase down all fossil fuels. That was the subject of intense wrangling but in the end,

it failed, and the resolution included was the same as that in Glasgow.

- **World Bank reform:** A growing number of developed and developing countries called for urgent changes to the World Bank and other publicly funded finance institutions, which they say have failed to provide the funding needed to help poor countries cut their GHG emissions and adapt to the impacts of the climate crisis. Reform of the kind widely discussed at COP27 could involve a recapitalization of the development banks to allow them to provide far more assistance to the developing world.
- **Adaptation:** Activities such as building flood defenses, preserving wetlands, and regrowing forests can help countries to become more resilient to the impacts of climate breakdown. But poor countries often struggle to gain funding for these efforts. Some funds were promised (but not fulfilled) for such adaptations. At Glasgow, countries agreed to double that proportion, but at COP27 some sought to remove that commitment. After some struggle, it was reaffirmed. However, the total funds promised still need to be received.
- **Tipping points:** A reference to the key finding of "tipping points" was put in being a warning that the climate does not warm in a gradual and linear fashion, but that we risk tripping feedback loops that will lead to rapidly escalating effects. These include the heating of the Amazon, which could turn the rainforest to savannah, transforming it from a carbon sink to a carbon source, and the melting of permafrost that releases the powerful GHG methane.

COP 28: *United Arab Emirates*

There was controversy from the beginning concerning remarks made by the COP28 president, who is also the CEO of the UAE's state-owned oil company. He stated that there was "no science" indicating that a phase-out of fossil fuels was needed to limit global warming to 1.5°C this being the goal of the Paris Agreement. He later clarified his views and said that a phase-out was "inevitable and essential."

- A further challenge was the lack of ambition and urgency shown by some of the major emitters, such as China, India, and Russia. They did not submit new or updated nationally determined contributions (NDCs) following the voluntary pledges to reduce emissions under the Paris Agreement. The current NDCs are not seen as being good enough to close the gap between the projected emissions and the level required to keep the global temperature rise below 1.5°C.
- There was much controversy about a pledge to move away from fossil fuel production. Long negotiations resulting in a pledge to phase-out fossil fuels were watered down to a decision to "transition away" from fossil fuels, which seems much weaker. All of this was a consequence of pressure from fossil-fuel-producing countries.
- An agreement was reached to create a fund to help the most vulnerable countries to deal with the costs of adapting to climate change with funding pledges being made by some countries. However, the amount pledged was a paltry $400 million with a tiny contribution from the United States. Given the scale of climate change damage, it will be a "drop in the ocean." Secondly, these amounts are just pledges made, and we all know from experience that there is great doubt that the funds actually received with bear any resemblance to the pledges.

Conclusions on COPs

Having read the above, the reader is free to draw their own conclusions about what COP has achieved. Clearly, a huge amount of time and resources have been, and are still being, committed to COPs and international climate change, and there have been some big achievements in relation to the Kyoto Protocol and the Glasgow Climate Pact. However, there are still many concerns which I suggest are as follows:

- Progress does not seem fast enough to deal with the problems being faced and the achievement of net zero by 2050. Many climate scientists argue that the COPs are not responding quickly enough to the overwhelming scientific evidence available.
- A lack of engagement by some countries. This was exacerbated by the initial withdrawal of the United States from the Paris agreement in 2018.
- A failure by countries to provide funds to meet certain pledges, which require finance.
- The pledges made are not legally binding, and there is no international monitoring of the implementation of actions to meet the pledges. There may or may not be monitoring arrangement within a country. Finally, there is no sanction on countries failing to meet their pledges.

The Scale of the Problem

Before considering what actions are being taken, it is important to first consider the nature of the problem being addressed. We have already seen that "Net Zero" requires countries to get to a situation where they have net zero GHG emissions based on the fact that the actual emissions they make will be balanced by removals from the atmosphere. If everyone achieves this, then the climate change problem should be solved. However, given that, currently, this is not the case, countries have two main options.

- Reduce emissions of GHG and/or
- Increase removals of GHG

To start with, let us have a look at where the emissions of GHG come from. If we start with CO_2, then the sources of emissions can be classified in a number of ways. Figure 6.1 illustrates sources of emissions by sector.

It can be seen that the three big emitting sectors, which are over 50 percent of total, are electricity and heat, transportation, and manufacture/construction.

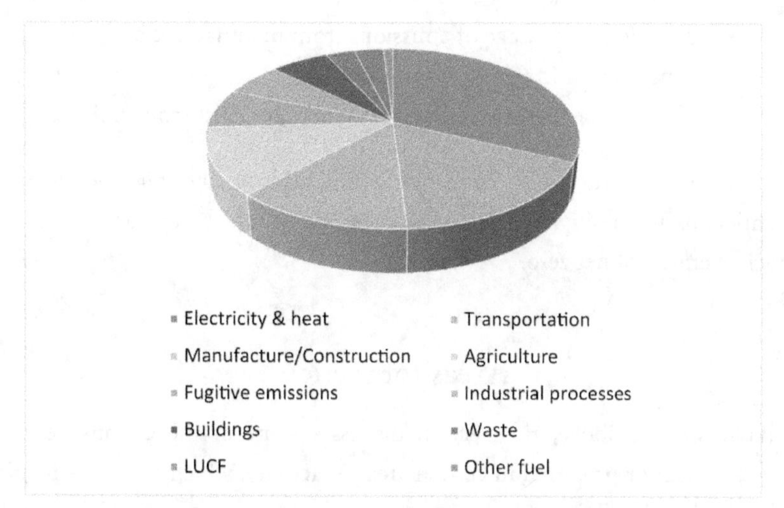

Figure 6.1 Percentage distribution of global GHG emissions by sector 2019

Another analysis of emissions is by country. I have not presented the individual data on emissions for all countries in the world, but this can be easily accessed. Instead, what I show in Table 6.1 is the composition of these emissions for a small number of critical countries that account for almost 70 percent of global GHG emissions.

A number of points should be noted:

- The above 10 countries account for almost 70 percent of global emissions with the remaining 30 percent arising from almost 200 other countries.
- The high percentage of emissions due to electricity production and heating across most countries.
- The high percentage of emissions derived from transportation in developed countries.
- The high proportion of emissions, termed fugitive emissions from Iran, Russia, and Saudi Arabia. These are due to the release of gases into the atmosphere consequent on extraction of oil and gas. Similar profiles will be seen for countries such as the Gulf states.

- The high percentage of emissions from manufacture and construction in China
- The high percentage of emissions from agriculture in India.

Clearly, if these 10 countries could address the high levels of emissions in the highlighted areas, then this would better facilitate the achievement of net zero.

Areas for Action

Based on the above, this section discusses some of the actions being undertaken or proposed in certain areas of activity, which generates high levels of GHG. These concern

- Electricity generation
- Transportation
- Manufacture and construction
- Fugitive emissions
- Air travel
- Land use change and forestry

A few comments will also be made about the issue of carbon pricing, which is a controversial area.

Electricity Generation

There are two basic approaches to reducing emissions relating to electricity generation:

- Reducing the use of fossil fuels for the generation of electricity and/or
- Using less electricity by one or more of the many means available such as insulation, lower room temperatures, reduced air conditioning, and so on.

Table 6.1 Analysis of GHG emissions by country

	China	US A	India	Russia	Japan	Iran	Canada	Germany	Saudi Arabia	South Korea
Total emissions (million tons)	12,129	5929	3377	1975	1167	901	779	754	747	697
	%	%	%	%	%	%	%	%	%	%
Electricity/heating	46	33	37	44	47	23	27	35	35	52
Manufacture and construction	23	7	17	14	16	11	9	12	12	9
Industry	10	4	5	3	6	4	3	3	14	12
Transport	8	31	9	13	18	16	25	22	19	16
Agriculture	5	6	21	5	2	4	7	8	1	2
Fugitive emissions	5	7	1	28	1	21	8	1	11	0
Buildings	4	9	5	11	9	14	10	16	1	7
Waste	2	2	2	6	1	5	2	1	4	1
Other fuel burning	1	1	2	1	1	1	2	1	0	1
Aviation and shipping	1	3	1	3	3	1	1	5	3	6
Land use change and forestry	-5	-4	-1	-28	-3	0	5	-4	0	-7
TOTAL	100	100	100	100	100	100	100	100	100	100

Source: Our world in data.

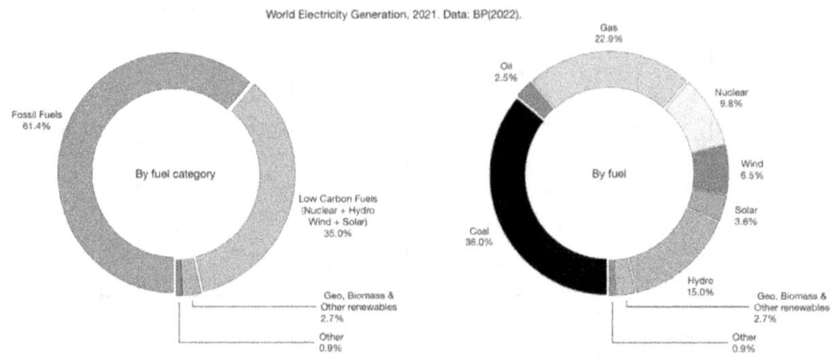

Figure 6.2 Sources of electricity generation

Globally, it can be seen in Figure 6.2 that fossil fuels dominate electricity generation. Although this will vary between countries, fossil fuel generation will usually be an important source of electricity.

Turning to the UK, throughout the 20th and the early 21st centuries, the bulk of the country's electricity was generated through the burning of fossil fuels. The contribution from fossil fuels has declined in recent years while electricity generation from renewable sources has increased significantly. To this can be added the contribution from nuclear. Using the UK as an example, we see that as of 2022, the main sources of electricity generation are as shown in Figure 6.3.

Figure 6.3 shows the huge changes which have taken place in UK electricity generation over a 10-year period with the decline of fossil fuels and the growth of renewables. This picture will be repeated in many other countries.

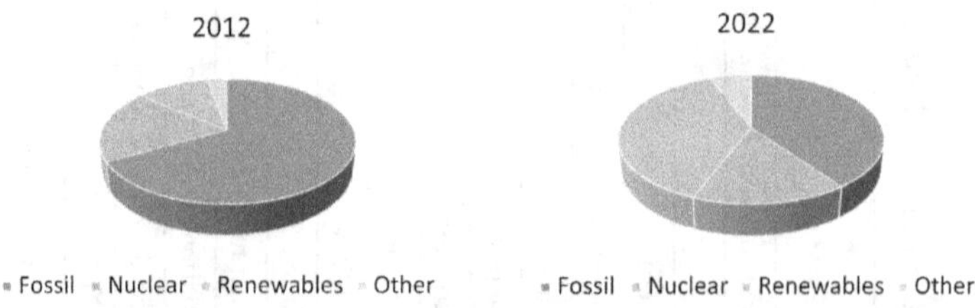

Figure 6.3 Percentage sources of UK electricity generation

Renewables

If we look at renewable sources of electricity generation, the following methods can be used:

- Wind
- Solar
- Hydro
- Tidal
- Geo and Biomass

Basically, there are two main problems with renewables:

- ***Uncertainty:*** With the exception of tidal power, all of the other renewable approaches have an uncertainty attached to them especially in temperate climate zones. The sun may not shine, the wind may not blow, and there may be insufficient water for hydro. Indeed, the changes caused by climate change itself may interfere with the availability of wind, water, or sun. Many countries have long coastline and tidal ranges, and I am always surprised that more emphasis is not placed on tidal power which is guaranteed. There would be significant environmental damage to coastal areas as a consequence of tidal power but there will also be damage to coastlines as a consequence of climate change itself.
- ***Environmental degradation:*** Renewables such as windmills and solar panels cause environmental problems in themselves. With solar panels, larger scale solar facilities can raise concerns about land degradation and habitat loss while significant amounts of water may be needed for cell manufacture and cooling when in operation. Hazardous materials may be used in cell manufacture, and some types of cells will contain toxic materials. Windmills are the ultimate in embedded costs and environmental destruction. Each weigh 1688 tons and contains huge amounts of concrete, steel, iron, fiberglass, and "hard to extract" rare Earth elements. These rare Earth elements need to be extracted from the ground with considerable land degradation and may be

mined in countries where child labor is used in the mining process. Thus, while the benefits to be derived from the use of renewable sources of electricity may outweigh the costs described above, it shouldn't be regarded as a "free" environmental good.

The thrust of energy policy, in the UK and many other countries, is to move further away from fossil fuels and toward renewable sources of energy. The UK government has published a plan to decarbonize electricity generation by the year 2035 with reliance on renewables sources of energy and some nuclear energy. While this is a laudable aim, there must be some doubt as to whether it is achievable, and governments may have to resort to more nuclear power, which is already happening in some countries.

Nuclear Generation

Electricity generation through nuclear fission was developed some 70 years ago and has subsequently spread around the world. Nuclear fission power stations can generate electricity without any carbon emissions but there are serious drawbacks most notably with regard to safety and the disposal of spent nuclear materials. Nuclear fission will continue to be a significant source of electrical power, but there will always be concerns about the high levels of capital costs involved and the problems of siting of new such power stations.

The other aspect of nuclear generation concerns nuclear fusion. This is a different process to fission and involves fusing (as opposed to splitting) hydrogen atoms with an outburst of energy. Nuclear fusion is the basis for the hydrogen bomb developed in the 1950s while controlled nuclear fusion has been discussed for six decades but progress has been slow. A breakthrough, in the USA, was recently announced but the construction of a working fusion reactor must be many decades away.

Transportation

Emissions relating to transportation derive from the use of petrol or diesel engines in land transportation by car, lorry, train, or bus, and so on. Many countries have announced plans to reduce or eliminate petrol/diesel vehicles. For example, in November 2020, the UK government announced its intention to enforce the following in relation to new vehicles.

- Step 1: This will see the phasing-out for the sale of new petrol and diesel cars and vans starting in 2030.
- Step 2 will see all new cars and vans be fully zero emission at the tailpipe from 2035.
- Between 2030 and 2035, new cars and vans can be sold if they have the capability to drive a significant distance with zero emissions (for example, plug-in hybrids or full hybrids), and this will be defined through consultation.

However, it should also be noted that some governments seem to be backsliding on these commitments and extending the timescale for eliminating petrol/diesel vehicles.

Ultimately, there may well come a time, at some unknown date in the future, when no new petrol or diesel cars will be allowed to be

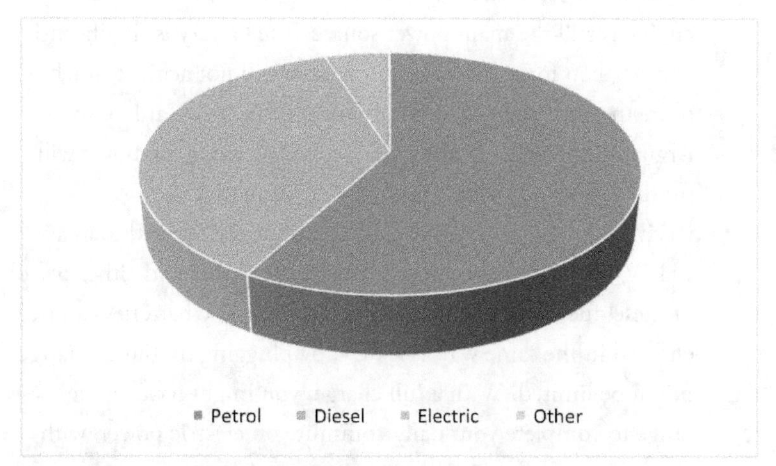

Figure 6.4 Cars on UK roads 2022 by fuel type

manufactured. However, it will not be illegal to drive a petrol/diesel car, and given the experience of classic cars, we can expect a number of petrol/diesel cars to continue on our roads for many years to come.

However, we can expect the total numbers of nonpetrol/diesel cars to decrease sharply. The two competing alternatives are electric vehicles and hydrogen-powered vehicles and these are considered in turn.

Electric Vehicles

Electrically powered vehicles are seen as being one of the key means of solving global warming. They do this by not having much in the way of GHG emissions. However, "electric vehicles" can come in several different types as listed below.

- **BEVs or EVs:** These acronyms basically mean the same thing standing for electric vehicle or battery-powered electric vehicle. BEVs stand out from most cars on the market in that they don't have internal combustion engines. Instead of petrol or diesel, these vehicles run solely on battery power with the battery being recharged as required.
- **HEVs:** Meaning hybrid electrical vehicles often referred to as hybrids, or referred self-charging hybrids. These are powered by a combination of electricity and a petrol or diesel engine. However, a HEV cannot be plugged into the mains, as the engine is still the main power source. The battery is significantly smaller than for an EV, and the vehicle will not achieve much more than a couple of miles of pure electric range at low speeds. Arguably, the HEV is not really an electric car at all, but it will produce less CO_2 than a petrol or diesel model.
- **PHEVs:** A plug-in hybrid vehicle, commonly referred to as a PHEV, uses batteries to power an electric motor and either petrol or diesel fuel to power an engine. However, the batteries can be charged in the same way as a BEV, by plugging in, but the range might be limited. With a full charge, you might have enough range to complete your daily commute on electric power, with

the petrol or diesel engine available should you need to travel further afield. A PHEV will typically start in electric mode and will run on electricity until the battery pack is depleted. You can also choose to save the electric range for urban use.

There are a huge range of statistics available about electric vehicles but the important point to note from table below is that electric cars of all types (including hybrids) are still a small proportion of the total. For the UK, this is shown in Figure 6.4.

The total numbers of vehicles has grown in recent years, with an increasing proportion of new registrations being electric cars. This trend is expected to continue in the future. In light of this, we need to consider what electric cars involve.

This book aims to present a realistic picture regarding climate change and the importance of guarding against the idea that technological developments are certain to occur and will solve the problem. At this point, the author will express a view that the hype around electric cars seems extreme, even by the standards of climate change. We will consider this under the following headings:

- ***Range:*** Depending on the type of car, the way it is driven and the terrain, the average petrol/diesel car might be capable of clocking up 500–550 miles on a full tank of petrol or diesel. With battery electric cars, the situation is quite different. Ignore the manufacturers' claims, which are inflated, and you will probably find the average electric car can do 250–300 miles on a full charge. Hybrids can achieve more. Manufacturers will claim that this will improve, and it will to some extent. However, it is difficult to see electric cars having the same range as petrol/diesel cars without having much larger batteries, which make the cars even heavier and more expensive to purchase. Also, battery range might decline with age. If you live in an urban area and most of your journeys are fairly short, then this might not be a problem. However, if you live in a rural area, have a long drive to work, or have family living hundreds of miles away, then (in the absence of decent public transport) you may have real problems in looking for somewhere to recharge away from home.

- ***Re-fueling (charging):*** There are two issues here—charging time and charger availability. Clearly, charging the battery of an electric car takes longer than filling an existing car with petrol. A trip to the pumps can take about 5 minutes, but at the moment, a rapid charger will charge an electric car in 20 to 40 minutes but that will vary according to a number of factors. It is suggested that charging times might reduce in time, but I doubt it will get anywhere near 5 minutes without having heavily polluting batteries. The next issue is charger availability. Those who have a house with a drive or garage will be able to install a home charger. However, in the UK, around 30 percent of people do not have a house with a drive or garage, and this percentage is much higher in urban areas. They and others who wish to charge away from home will have to rely on public chargers, which may or may not be proximate to where they live.
- ***Infrastructure:*** Currently, most people live in an area where they can get to a petrol station and top-up their car without having to queue for a considerable time. With electric cars, in addition to charging at home, there will need to be a large number of public charging points. These charging points will be found at a wide variety of locations including supermarkets, shopping centers, public car parks, hotels, and service stations alongside petrol pumps. There will probably be a need for different charging options as there already are with petrol. These charging facilities will require a complex infrastructure including charge points, cabling, transformers, and so on. The UK government has set ambitious targets for 300,000 public chargers to be operational by the time of the ban on new petrol and diesel vehicles, but at current growth rates, this looks set to fall far short of that. Government figures showed that in the last 3 months of 2022, an additional 2418 chargers were installed around the UK. However, to meet the target of 300,000 chargers by 2030, the country needs almost four times as many

chargers to be installed every quarter. Furthermore, some will argue that 300,000 chargers are just not sufficient. As already noted, the lack of chargers will lead to long waiting times and queues. While it is expected that the electrical power for recharging batteries will come from renewable sources, there are also significant concerns that the power distribution networks will not be sufficiently robust to cope.

- ***Costs:*** The costs associated with vehicles concern: purchase cost, lifespan, running costs, and external costs such as insurance, taxation, and so on. This is a difficult area. Electric cars are currently more expensive to purchase than petrol or diesel cars, but this may reduce in time. Running costs in the longer term are largely unknown because of lack of experience with regard to costs of repairs and servicing. At the moment, fuel costs for an electric car are less than for petrol/diesel, but this may change to cost changes and/or changes in government taxes on fuels.
- ***Emissions:*** With BEVs, the situation is clear cut, and there are zero direct emissions from vehicles. With hybrid vehicles, there will be some direct emissions, but these can be kept low if most journeys are short local ones which can be done using battery power. However, I don't think this is the end of the story on emissions. It is estimated that if all car and taxi transport were by electric vehicles, the total amount of electricity needed by the country would rise by approximately 20 percent. It is anticipated that this additional electricity will be generated by renewable means, and it is estimated that annual electricity demand from electric cars will be around two-thirds of the amount of additional power produced by the offshore wind generators planned to be installed by 2030. This looks like a rather vulnerable situation, given the other demands on renewable electricity.
- ***Rare Earths:*** Another important point that needs to be emphasized is that each car battery produced requires the use of significant amounts of rare earth elements such as lanthanum, cerium, praseodymium, neodymium, promethium, samarium,

europium, gadolinium, terbium, dysprosium, holmium, erbium, thulium, ytterbium, and lutetium. These metals are used in electric car batteries because of their ability to store and release electrical energy. However, there is a downside. These rare earth elements are often found in countries that are poor, unstable, and have ineffective labor laws. Such countries may also resort to political blackmail regarding supplies of these elements. Also, mining can involve huge environmental degradation. A project is currently being considered to mine these rare earth elements from the seabed of the Pacific Ocean, with potentially huge and damaging impacts on the ecosystem. Another potential solution on the horizon is the development of sodium batteries since sodium is freely available as sodium chloride (salt) in sea water. However, one of the major disadvantages of sodium batteries is their relatively low energy density, meaning the amount of energy stored relative to the battery's volume. Lower energy density means bulkier and heavier batteries, and this could problematic.

Hydrogen-Powered Vehicles

Hydrogen-powered vehicles work quite differently to electric vehicles. While electric vehicles use electrical energy that has already been generated elsewhere (and stored away in the vehicle's the battery pack), hydrogen vehicles generate their own electrical energy from a hydrogen store (via fuel cells). Hence, hydrogen fuel cells vehicles are, in essence, a hybrid vehicle but with fuel cells (converting hydrogen) instead of an internal combustion engine (converting gasoline). The development of hydrogen-powered vehicles has lagged behind that of electrical vehicles but might be now catching up. However, in considering hydrogen-powered vehicles, the following points are important.

- ***Range:*** As already mentioned, the range of any vehicle will vary according to a number of factors. However, it is probably fair to say that hydrogen-powered cars will have a range comparable with petrol/diesel cars and significantly higher than electrically powered battery cars.

- ***Refueling and infrastructure:*** Refueling a hydrogen car at a pump takes 5 minutes or less, much like refueling with petrol/diesel. This compares to electric cars, which can take 6–8 hours for a level 2 charger (the most common type currently), down to 30 minutes for the very fastest (and rarer) level 3/4 charger. Recharge times for electric cars probably will fall, but almost certainly won't be as quick as the 5 minutes of hydrogen. Another important issue concerns the availability of hydrogen refueling. The creation of a fueling station network is essential to the market development of these new vehicles, and there is much uncertainty surrounding this.
- ***Costs:*** Overall, hydrogen cars are probably comparable in cost to petrol-powered vehicles, but both will have running costs higher than electric cars. However, one cannot overemphasize the uncertainty in this area and things may radically change.
- ***Safety:*** The two main hazards from fuel cell and hydrogen-powered vehicles are electric shock from the high voltages being used and fuel flammability. Clearly, there will be a variety of national and international safety standards, but the key questions are whether they will be strict enough and the serious consequences of safety failure.
- ***Emissions:*** The main emission from hydrogen-powered vehicles is water vapor, derived from the fusion of hydrogen and oxygen. As noted in Chapter 5, while water vapor is not a GHG by the strict definition, scientists now believed that it does act as a catalyst that increases the potency of other GHGs such as CO^2 and methane.

Overall, the question of hydrogen or rechargeable batteries will probably come down to individual choice and circumstances. This will be influenced by whether you live in town or country, whether you make short journeys or long journeys, whether you can recharge your car at home, and so on. Moreover, there is a huge amount of uncertainty about a variety of issues, and the wise choice might be to hold off the decision until things become clearer.

A final issue is purchasing cost. At the moment, the prices of electric cars are significantly higher than the equivalent petrol car. This may explain the surging demand and prices for used petrol/diesel cars. It is often assumed that prices for electric cars will decline with increased production numbers, but care is needed. The batteries are a significant proportion of cost, and as noted above, these require rare earth elements which are expensive and may increase in price substantially. It remains to be seen whether electric car prices align with the prices charged for equivalent petrol/diesel cars and to what extent they will be affordable for some parts of the population.

Manufacturing and Construction

Manufacturing, especially those using the cheap construction staples of steel and cement, accounts for a large proportion of GHG emissions. That puts this sector as having the same order of pollution as power or transportation sectors, which receive far more attention in terms of policies and investments. However, the basic problem is that there is currently no way to make steel or cement without releasing climate-warming emissions from steel and cement-making since both are processes that release CO_2 as a by-product. Also, the manufacturing and construction sector is set to grow as the global population climbs and countries further develop their economies. In addition, consequent on climate change, there will be a need for some form of new housing in many developed and developing countries since shelter must be seen as one of the basic needs of humanity.

While research into alternative approaches is being undertaken, no obvious solution has yet emerged and so these fall into the category I always emphasize of relying on future scientific and technological solutions to this problem. There just may not be an easy solution.

Fugitive Emissions

Fugitive emissions are gases and vapors accidentally released into the atmosphere. Such fugitive emissions like methane are unintentionally released by the oil and gas industry and add a GHG to the atmosphere that's over 25 times stronger than CO_2. In the table shown earlier, it can be seen that high levels of fugitive emissions derive from Iran and Russia. Similar profiles will be seen in other oil-producing countries, like Saudi Arabia and the Gulf States. Other fugitive emissions come from industrial activities, like factory operations. All these emissions contribute to climate change and air pollution.

Although these emissions are stated as being unintentional, in the case of gas emissions, the companies involved are wasting a product, which could be sold for revenue. However, it is possible that some of these emissions are caused by lack of care in production operations. Such fugitive emissions can be reduced by such actions as replacing old, outdated valves, making sure valves are installed correctly, having proper maintenance programs, and so on. While fugitive emissions are a fairly small proportion (5.9%) of global GHG emissions, they are significant in a fairly small number of countries. As it seems fairly easy to counter these emissions, this should be seen as a priority area for action.

Air Travel

Air travel contributes a fairly small amount of global GHG emissions (1.9%) but is easily the most polluting form of travel as shown in Table 6.2, which uses a domestic journey as an example.

Passenger numbers on aircraft rose from 2.7 billion in 2010 to 4.5 billion in 2019, and there were expectations that passenger numbers

Table 6.2 GHG pollution levels of different modes of travel

Travel mode	Grams of carbon dioxide equivalents per passenger-kilometre
Ferry	19
Rail	41
Bus	105
Medium car (petrol)	192
Domestic flight	255

Source: https://ourworldindata.org/travel-carbon-footprint.

would increase from then on by around 4 percent every year with the numbers of aircraft doubling by 2038. However, these forecasters didn't know about the Covid19 pandemic.

In 2020 with the start of Covid, passengers plummeted to 1.8 billion and have yet to recover to pre-Covid levels. Hence, the sorts of growth in passenger numbers talked about in 2019 might now seem fanciful (The Independent 2019). Time will tell. Similar trends too place in relation to freight.

However, aviation remains one of the hardest sectors to decarbonize. Nothing propels a commercial aircraft as efficiently and economically as fossil fuel. Thus, there is only a limited amount that can be done to reduce harmful emissions from existing aircraft engines, so, as with land-based vehicles, aircraft companies are looking at more environmental-friendly approaches to air transport. The two candidates, once again, are electric planes and hydrogen powered planes.

There are, however, some major problems with both of these solutions.

Electric planes will require large batteries to power their engines, and such batteries are very heavy and provide for only a limited flying range. While domestic flights may prove possible through electric planes, it will require some huge breakthroughs in battery technology for long haul to become feasible.

When we turn to hydrogen-powered planes, we see some huge challenges for aircraft architecture and design (specifically in areas of fuselage redesign and hydrogen storage integration) to issues surrounding the hydrogen supply chain. Liquid hydrogen, which is easier to store onboard than gas, has to be kept at -253°C or it boils off. The tanks to contain it are not only heavier but four times the size of conventional fuel storage. This imposes constraints on the range and capacity that commercial aviation may struggle to accept. It will probably be several decades before hydrogen-powered planes become viable and then they may be restricted to short haul journeys only. I repeat my warning that just because technological solutions are needed doesn't mean they will automatically happen.

In this situation, the only real option for reducing carbon emissions from air travel is to reduce the numbers of people who take flights.

This could be done by stiff taxation of airline tickets to reduce demand especially in relation to nonessential traffic such as long-haul tourist flights. As well as reducing carbon emissions, this would have the added advantage of reducing environmental damage being caused to popular tourist destinations such as Machu Piccu, Himalayan trails, and so on.

Land Use Change and Forestry

This is an unusual sector in that it combines releases of GHG into the atmosphere but also removes (sequesters) GHG as shown in Figure 6.5.

Thus, it will be seen that Figure 6.5 shows a negative figure, meaning that sequestrations are greater than emissions. Hence, the mitigation of climate change requires actions to reduce emissions and increase sequestrations. There are many actions which could be taken, and just a few examples are shown below:

- Adoption of sustainable and low-carbon crop farming practices.
- Protection, maintenance, restoration, and expansion of woodlands which would provide substantial carbon storage.
- Protection and expansion of wetlands and peatlands.
- Reductions in deforestation and crop burning.
- More sustainable livestock farming to reduce methane emissions.

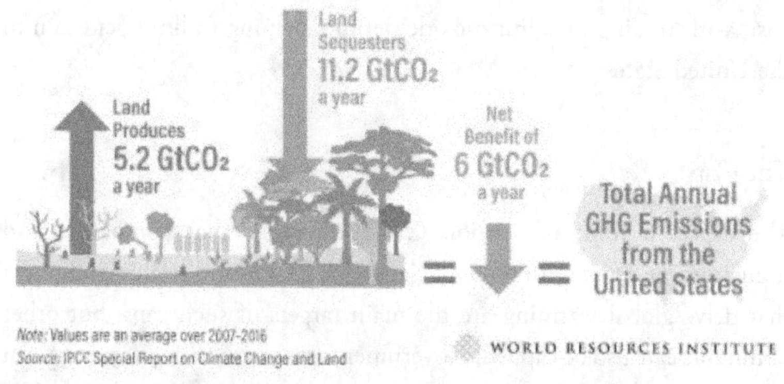

Figure 6.5 Land emission and sequestration

- A shift from animal to plant diets with a high potential for reducing carbon footprints.
- Recycling systems for bio-degradable waste enabling food waste to be composted or used for bioenergy production and so on.

Carbon Pricing

Carbon pricing is an instrument that captures the external costs of GHG emissions, this being the economic costs of emissions to the public in terms of such things as damage to crops, healthcare costs from heat waves and droughts, and loss of property from flooding and sea level rise, and so on. A price on carbon helps shift the burden for the damage from GHG emissions back to those who are responsible for it and who can avoid it. Instead of dictating who should reduce emissions where and how, a carbon price provides an economic signal to emitters and allows them to decide to either transform their activities and lower their emissions or continue emitting and paying for their emissions.

There are various ways of using carbon pricing, and carbon trading is one of them. Many governments around the world are now using carbon pricing in such a trading arrangement. It involves a complex mechanism designed to put downward pressure on GHG emissions from companies while giving those companies some flexibility in how they go about reducing their emissions. It works by operating a market-based system for trading in carbon. An alternative name that better describes the approach is cap and trade, and such schemes have previously been very effective in tackling environmental problems in the past, with trading in sulfur dioxide permits helping to limit acid rain in the United States.

The Cap

The cap is a limit on carbon GHG emissions that companies and industries can incur with no financial penalty. CO_2 and related pollutants that drive global warming are the main targets of such caps, but other pollutants can also be capped. Governments set a carbon cap across a given industry, or ideally the whole economy, and the cap typically declines over time, providing a growing incentive for industry and businesses to reduce their emissions while keeping production costs down. The total amount of the cap is split into allowances, each permitting a company to emit a certain amount of GHGs. The government distributes the allowances to the companies, either for free or through an auction.

Trade

Having received their carbon emissions quota, companies can trade with one other to either purchase or sell carbon credits at the regulated carbon price. If a company curbs its own carbon emissions significantly, it can sell the excess permits on the carbon market for cash or keep them for future use. If it's not able to limit its own emissions within its own cap, it may have to buy extra permits from elsewhere.

There are many debates about the merits of carbon trading mechanisms, but the big attraction for governments concerned with stemming CO_2 is that carbon trading is probably much easier to implement than expensive direct regulations or administratively difficult carbon taxes. Others argue that carbon trading is a distraction from the main emphasis of reducing GHG emissions and that it is often a mechanism for large, developed nations to offload their GHG emissions on to poorer nations.

Conclusion

At the start of this chapter, I discussed the concept of net zero climate emissions and discussed why this is now seen as the key objective of the world in terms of preventing calamitous climate change. I also argued that actions to prevent climate change could be taken at all levels from the global to local communities.

Climate change is an international problem which requires internationally coordinated solutions. Therefore, in this chapter, I spent time analyzing and summarizing the outcomes of the 28 COP events, which have taken place since 1995. Although there is a huge amount of scientific evidence and advice given to the COPs, the reality is that it is very much a political forum rather than a scientific one. Hence, decisions are made on the basis of negotiations and trade-offs between countries. Not surprisingly, the outcomes of the COPs tend to have been nonoptimal and inadequate to the challenges being faced. One does wonder, however, whether the delegates to the COPs have a handle on reality about what is going to happen. I will return to this in a later chapter.

I noted earlier that under the Paris Agreement, countries agreed to make plans to limit their emissions of GHGs. This agreement clearly defines 2.0°C as the upper limit for global warming but also lists 1.5°C as a more desirable goal because it reduces the risk of the worst outcomes of climate change in most of the world. Does a difference of 0.5°C really matter? The evidence shows that it definitely does matter and a 1.5°C rise is to be strongly preferred.

While the growth of global emissions has slowed in recent years, there is still a large and growing gap between current commitments and what would be needed to avoid exceeding these global temperature limits. In light of this, one analysis (Carbon Brief 2020) suggests the following:

- The world will likely exceed 1.5°C between 2026 and 2042 in scenarios where emissions are not rapidly reduced, with a central estimate of between 2030 and 2032.
- The 2.0°C threshold will likely be exceeded between 2034 and 2052 in the highest emissions scenario, with a median year of 2043.
- In a scenario of modest mitigation—where emissions remain close to current levels—the 2.0°C threshold would be exceeded between 2038 and 2072, with a median of 2052.

Some recent evidence has suggested that the situation might be improving. A report from the International Energy Agency (IEA) suggests that global carbon emissions will peak in 2025 (and will then decline) thanks to massively increased government spending on clean fuels in response to Russia's invasion of Ukraine (The Economic Times 2022). However, the IEA analysis also suggested that current government policies would still lead to global temperatures rising by 2.5°C, which would have catastrophic climate impacts. Thus, it seems that even taking account of everything that is taking place in the world, it still seems likely that global temperatures will rise by at least 2.0°C. This is not a good news.

In the latter part of this chapter, I considered what is happening or promised in the key areas of action to achieve net zero by 2050, these being: electricity generation, transportation, manufacture, construction, fugitive emissions, and air travel. There are many potential difficulties ahead in all of these areas these being technical, financial, and political. There is a lot of discussion and even hype about what is happening in these key areas but there are also concerns being expressed about the rate of progress and the achievability of what is being proposed. The key problem is that it is difficult to be certain about the extent of what is promised will actually be implemented in time to achieve net zero by 2050.

Overall, the conclusion of this chapter must be that the situation is not promising and may be alarming. There is a general consensus that we will pass1.5°C of global warming in the next few years and we may even be heading toward a breach of 2.0°C and beyond with major consequences for the planet and its inhabitants. If the pledges made by various countries, as part of the COP process, go seriously astray because of other factors such as wars, economic downturns, political turmoil, and so on, then the impacts don't bear thinking about.

CHAPTER 7

How Are Countries Likely to Respond to Climate Change?

Introduction

On 20 March 2023, the Intergovernmental Panel on Climate Change (IPCC), made up of the world's leading climate scientists, set out the final part of its mammoth sixth assessment report. The report and comments on it were full of stark and urgent language such as:

- *"A 'final warning' on the climate crisis, as rising GHG emissions push the world to the brink of irrevocable damage that only swift and drastic action can avert"*
- *"Only swift and drastic action can avert irrevocable damage to world"*
- *"This report is definitely a final warning on 1.5 °C"*
- *"Humanity at the climate crossroads: highway to hell or a liveable future"*

This report outlines the dangers to the planet and humanity, and points out that we do have the means to take the necessary actions needed. It is a matter of having the will to do so. In light of this, one wonders why governments are not falling over themselves to take measures to mitigate climate change before it is too late. There are clearly pressures for governments to take the urgent actions the report recommends, but the fact that it is not happening suggests that there are also pressures not to take actions that are too radical. In this chapter, I tried to explore what those pressures are and what are the implications.

My focus in this chapter is on the 10 countries which are the largest emitters of GHG and which, taken together, contribute to about 70 percent of all emissions. This chapter comprises the following sections.

- Country vulnerability to, and readiness for, climate change
- Country features
- Possible country responses to climate change
- Collaboration and trust between countries

Vulnerability to, and Readiness for, Climate Change

At the outset, it is probably fair to say that individual countries when considering the implementation of climate change mitigation measures have to keep two factors in mind.

- The impact of these measures on their own country—its people, its economy, its environment, its public finances, and so on.
- The impact of the measures internationally and, in particular, the impact on global warming. There is also a political factor here since the impact of a small country on global climate change would be negligible, but every country must be seen to doing their bit.

Table 7.1 provides some interesting information. It shows the 10 most polluting countries in the world (in terms of GHG emissions) ranked in order. For each of those countries, it shows two other things, namely, a country's vulnerability to climate change and its readiness for dealing with climate change. For comparison, the table also shows the data for three countries which are relatively small polluters.

There are four points to be emphasized from this table.

1. Aside from India, the countries which are the largest polluters are generally at the lower end of vulnerability to climate change.
2. Aside from India and Iran, the countries which are the largest polluters are generally at the upper end of readiness for dealing with climate change. However, readiness here is comparative and, as we will see later, most countries are not very ready at all.

Table 7.1 Most and least polluting countries

Country	Global ranking for GHG emissions (1-192)	Vulnerability to climate change *1 = Least vulnerable* *192 = Most vulnerable*	*Readiness for climate change* *1 = Most ready* *192 = Least ready*
Most polluting countries			
China	1	68	36
USA	2	23	18
India	3	132	104
Russia	4	28	37
Japan	5	49	11
Iran	6	57	102
Canada	7	8	19
Germany	8	4	11
Saudi Arabia	9	74	39
South Korea	10	52	19
Least polluting countries			
Somalia	165	181	120
Liberia	171	177	167
Central African Republic	179	175	192

Source: Adapted from ND-Gain (https://gain.nd.edu/our-work/country-index/rankings/).

3. The three countries which are among the lowest polluters in the world are all quite vulnerable to climate change and low in terms of readiness for dealing with climate change.
4. The 10 top polluting countries are all high- or middle-income countries. Hence, they have the financial resources needed to prepare for climate change adaptation. These financial resources are not available to low-income countries and, indeed, much of the debate in the COP summits concerned the provision of such resources to poor countries.

Aside from the above, another source of research suggests that while the top 10 polluting countries may be better prepared for climate change than poorer and less polluting countries, this does not mean they are well prepared. For each country, an independent research source (CAT) provides its analysis of the sufficiency of actions being taken to mitigate climate change with respect to the globally agreed aims of holding warming well below 2.0°C, and pursuing efforts to limit warming to 1.5°C. The organization provides recent detailed assessments of a number of factors but then provides an overall headline judgment about the sufficiency of the actions of each country. For the top 10 polluting countries, the results of this analysis are as shown in Figure 7.1.

Even allowing for some degree of imprecision, and possibly exaggeration, these results present a stark picture of lack of preparedness.

At this point, the story of Easter Island is something to think about. Easter Island is a small island in the South-eastern Pacific. It was covered with palm trees for over 30,000 years but is treeless today. It was the home to a thriving community with several thousands of people living good lives and with a rich culture. One by one, though, the trees on this isolated island were cut down. They were cut down for housing, fuel, or to make

Country	Degree of sufficiency	Date Assessed
China	Highly insufficient	November 2022
USA	Insufficient	August 2022
India	Highly insufficient	November 2022
Russia	Critically Insufficient	November 2022
Japan	Insufficient	October 2022
Iran	Critically Insufficient	September 2021
Canada	Highly Insufficient	December 2022
Germany	Insufficient	November 2022
Saudi Arabia	Highly insufficient	November 2021
South Korea	Highly insufficient	March 2022

Figure 7.1 Sufficiency of action on climate change

Source: Climate Action Tracker (https://climateactiontracker.org/).

tools, or boats. And finally, the last tree was gone and eventually so were the vast majority of its people as there was no more wood to build or make anything. The key question posed by anthropologists, psychologists, and so on was "What did the man on Easter Island think or say before he chopped down the last tree" which ensured disaster for the island. Doesn't this remind you of climate change?

Country Features

Overview

Although the numbers alter frequently, and are sometimes disputed, the world today can be regarded as comprising 197 countries in six continents of which

- 56 countries are in Africa
- 48 in Asia
- 44 in Europe
- 33 in Latin America and the Caribbean
- 14 in Oceania
- 2 in Northern America

Worldometers

Also, to be noted is the continent of Antarctica which is of great importance to climate change. Antarctica, which has no permanent residents, is governed by about 30 countries, all of which are parties of the 1959 Antarctic Treaty System.

Now, as is frequently mentioned in this book and elsewhere, climate change is an issue which is international in nature, and which must be addressed by all nations working collaboratively. Nevertheless, strong progress on these large and complex issues will be determined by the attitudes and policies of the individual countries concerned.

Of particular importance are the 10 countries who (as mentioned above) contribute 70 percent of total GHG emissions. It is these countries which I focus on in this chapter.

Table 7.2 Country populations

Country	Population 2022 (million)
China	1426
USA	338
India	1417
Russia	144
Japan	124
Iran	88
Canada	38
Germany	84
Saudi Arabia	36
South Korea	52

Source: Population Pyramid: www.populationpyramid.net/population-size-per-country/2022/.

This chapter takes a look at various features of these 10 countries which might influence their attitude toward climate change collaboration. There are three factors to consider.

- Country configurations
- Country economic structures
- Country inequalities

Country Configurations

Under this heading, I discuss three important factors.

- Population
- Geography
- Political structures

Population

The populations of countries in the world vary from 1.4 billion in China to less than a thousand in the Vatican City. Shown in Table 7.2 are the populations of those countries which are the 10 largest emitters of GHG in the world and therefore will hugely influence the pace of climate change.

It can be seen from the above that it is not just large population countries like China and the USA who are large emitters of GHG but also a number of much smaller countries, in population terms, such as Canada and Saudi Arabia.

Another issue to highlight is future trends in population numbers. This is illustrated in Figure 7.2.

We have already seen in an earlier chapter that the global population is projected to continue rising until the end of this century but to then start declining. However, such projections must be treated with caution due to the uncertainty about matters like fertility rates in certain countries. However, what Figure 7.2 shows is that underneath these global projections, we see huge variations in individual countries. Countries like India and the USA are projected to have huge increases in population numbers while countries like China, Russia, and Japan are projecting massive decreases in population numbers. Going beyond climate change, it is clear that if these population projections really come to fruition (which seems probable), it will have huge implications for these countries in social, economic, and political terms to an extent which can barely be speculated upon at the present time.

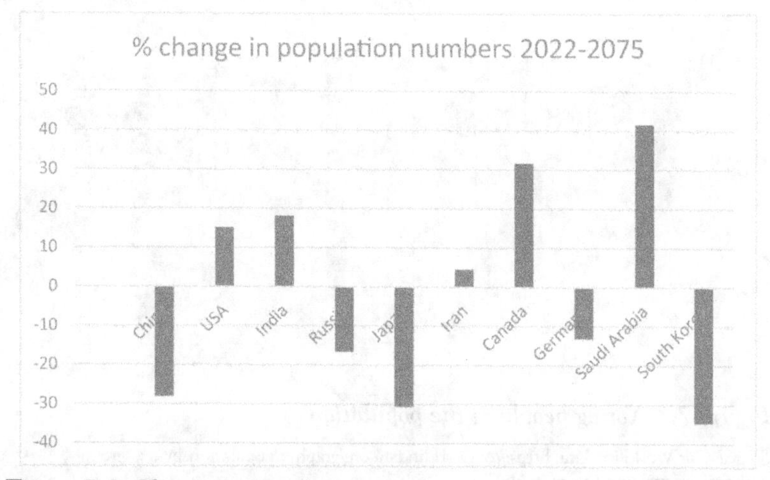

Figure 7.2 Change in population numbers

A third issue concerns the structure of the population and, in particular, the age structure. The age structure of a population is the distribution of people of various ages. It is a useful tool for economists, social scientists, public health and healthcare experts, policy analysts, and policy makers, and so on, because it illustrates population trends like rates of births and deaths. There are a host of social and economic implications in society consequent on the age structure and changes, like prospects for economic growth, labor supply, the resources that must be allocated for childcare, schooling, and health care, and so on.

The population of a country can be divided in many ways, but one approach is to segment it into three age-related groups.

- **Aged 65+:** This approximately represents those who have mostly retired from work and are no longer economically active and contributing to the country's GDP. However, this group is likely to be large consumers of state-funded health and social care.
- **Aged 0–14:** This group represents children who, in general terms, are not economically active. However, this group is also

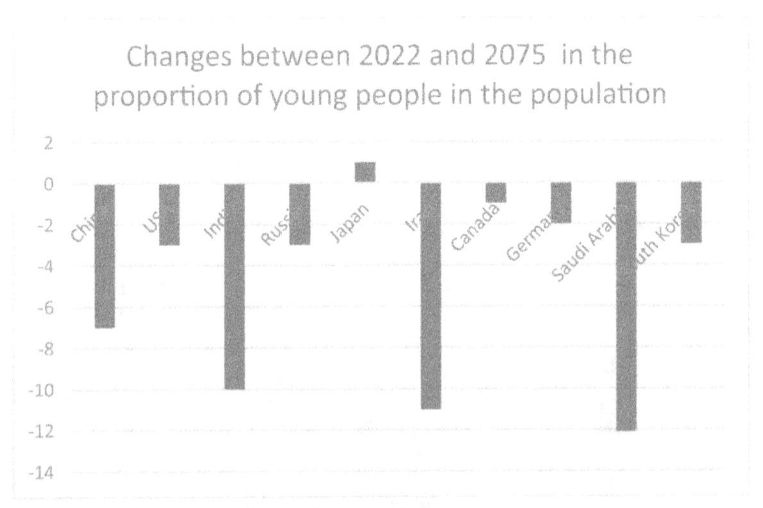

Figure 7.3 Young people in the population

Source: Our World in Data. https://ourworldindata.org/grapher/population-by-age-group-withprojections?country=~CHN.

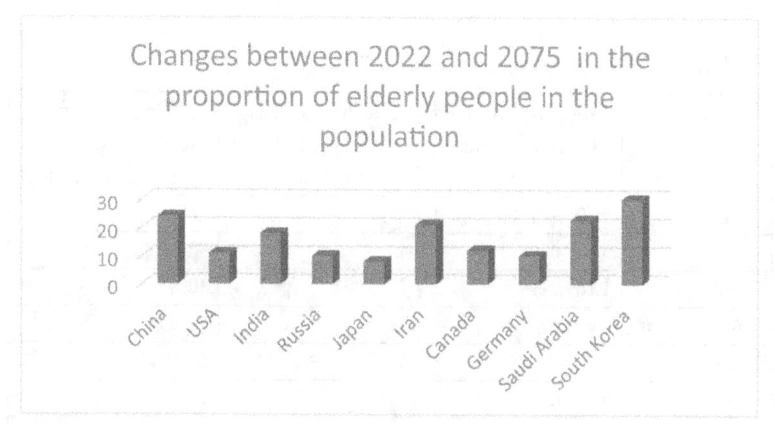

Figure 7.4 Elderly people in the population

Source: Our World in Data. https://ourworldindata.org/grapher/population-by-age-group-withprojections?country=~CHN.

likely to be large consumers of state-funded health care and education services.

- **Aged 15–64:** This group predominantly represents the working population of a country, which generates the GDP of the economy. This group probably consumes a lower level of publicly funded services than the other two groups

Currently, there are considerable variations between countries, in the age structure of their populations. However, the interesting issue is how population structures in various countries are expected to change over the next 50 tears. Figure 7.3 and Figure 7.4 illustrate the changes.

Two things can be, clearly, seen from these figures.

- The growth in the elderly segment of the population in all countries, albeit the growth is much higher in some than others.
- The decline in most countries in the young people segment of the population.

It will suffice to say that these changes will have profound social and economic implications for the countries involved.

Table 7.3 Land areas of countries

	Geographic area (square miles)	Global ranking of geographic area	Population density (persons per square mile)
China	3,705,407	3	388
USA	3,677,649	4	90
India	1,269,219	7	1087
Russia	6,601,670	1	22
Japan	145,937	62	870
Iran	636,372	19	132
Canada	3,855,100	2	10
Germany	137,882	63	609
Saudi Arabia	830,000	13	41
South Korea	38,690	107	1318

Source: Nationmaster. www.nationmaster.com/country-info/stats/Geography/Land-area/Square-miles.

Geography

The first aspect of geography concerns land area, and Table 7.3 illustrates the variations.

Among the most polluting countries, we see big variations in both land area and population density. However, if we look beyond the numbers, we see that the distribution and concentration of populations within a country will vary greatly from country to country. This is summarized in Figure 7.5.

Political Structures

Political science identifies a small number of models of government that exist in the world such as democracies, monarchies, autocracies, and so on. In practice, all of these models can be broken down into more specific submodels. All of these models (and others) can be found, to varying degrees, in the countries recognized by the United Nations.

A thrust of this book is that one of the key factors that will impact on the success or failure of global climate change is the governmental arrangements of the countries involved and the interplay between them.

Country	Population concentrations
China	• The world's most populous country. • Overall population density is less than that of many other countries in Asia and Europe. • Overwhelming majority of the population is found in the eastern half of the country. • The west of the country with its vast mountainous and desert areas, remains sparsely populated.
USA	• Large urban clusters are spread throughout the eastern half of the US and the western tier states. • Mountainous areas, deserts in the southwest, the dense boreal forests in the extreme north, and the central prairie states are less densely populated.
India	• Aside from the deserts in the northwest and the mountain fringe in the north, a high population density exists throughout the country. • The core of the population is in the north along the banks of the Ganges, with other river valleys and southern coastal areas also having large population concentrations
Russia	• Approximately, 68 percent of people live in the European part of the country, which makes up only 20 percent of the whole territory. • The rest of the vast area of land accommodates the remaining 32 percent of the population. • The regions of the Far North and the areas with severe climatic conditions, are especially poorly populated
Japan	• All regions of high population density lie on the coast. • One-third of the population resides in and around Tokyo on the central plain. • Over 90% of the population concentrates in big cities on the Pacific shore of Honshu. There are 14 cities with populations exceeding a million and another 200 have at least 100,000 residents. • Only 8% of Japanese live in rural areas, decreasing by 13% since 2000
Iran	• The population is concentrated in the north, northwest, and west, around mountain ranges. • The vast dry areas in the centre and eastern parts of the country, have a much lower population density
Canada	• Vast majority positioned in a discontinuous band within approximately 300 km of the US border. • 90% live within 100 miles of the US border. • Most populated province is Ontario, followed by Quebec and British Columbia. • The vast bulk of the area of Canada remains uninhabited or uninhabitable because of the challenging climate
Germany	• Fairly even distribution of population throughout most of the country, • Urban areas attracting larger and denser populations, particularly in the industrial state of North Rhine-Westphalia
Saudi Arabia	• Historically the population was mostly nomadic or semi-nomadic. • More settled since petroleum was discovered in the 1930s. • Most of the economic activities and the country's population is now concentrated in a wide area across the middle of the peninsula.
South Korea	• 70% of the country is mountainous. • Population is primarily concentrated in the lowland areas, where population density is quite high. • Gyeonggi Province which encompasses surrounds the capital of Seoul and the port of Incheon, is the most densely populated

Figure 7.5 Summary of country population concentrations

One aspect of this is the extent to which the government arrangements in a particular country may inhibit its ability to implement its agreed climate change policies. However, in dealing with the problems of global climate change, countries must also interface with one another. These relationships between countries, which will, to a great extent, be influenced by the respective governmental arrangements, will vary from hostile through to warm. In turn, this will have implications for the success of climate change policies.

Government structures vary enormously across the world but can be classified into a number of different systems including the following:

- **Democracies:** A democracy is a political system that allows for each individual to participate in the political process. There are several types of democracy where the defining characteristic is some level of citizen participation in the political system. Direct democracy is where citizens can directly participate in the governing process and the process of making laws via referenda. Switzerland is an example of this. More common is representative democracy whereby citizens elect representatives who actually make the laws. Features of a true democracy are that people are allowed to vote in secret and without coercion, there is a free media and political parties are free to operate without government interference.
- **Fake democracies:** These are countries that claim to be democratic and to actually hold elections. However, the elections held are not free and fair because people are intimidated to vote in a particular way; the media is not free and opposition parties are banned or interfered with. A good example of a fake democracy is Russia.
- **Totalitarian:** In this situation, there is no pretence at democracy and normally, a dictator is the main individual ruling the country. While there are lackeys and others who work for the dictator, he/she makes most of the decisions and usually has enforcers. The governed are usually not consulted in any way. One of the more common types of dictatorship is the military dictatorship, in which a military organization runs the political system. Alternatively, the military just exerts a great deal of pressure on the government, ostensibly running the country. In many cases, very few benefit from the decisions made in a dictatorship.

- **Theocracies:** A theocracy involves government by divine guidance or by officials who are regarded as divinely guided. In many theocracies, government leaders are members of the clergy, and the state's legal system is based on religious law.
- **Absolute monarchies:** An absolute monarchy is a form of government in which a single person, usually a king or queen, holds absolute, autocratic power. In absolute monarchies, the succession of power is typically hereditary, with the throne passing among members of a ruling family. Arising during the Middle Ages, absolute monarchy prevailed in much of western Europe by the 16th century but is now only found in a few parts of the world. It should not be confused with a constitutional monarchy where the monarch just has a constitutional or ceremonial role with political power resting with elected governments.

If we took our 10 most polluting countries in the world, the situation regarding political arrangements is as shown in Figure 7.6.

Thus, we see that six of the countries fall into the democracy basket while the other four fit into different categories. However, these categories can have some overlap since, for example, Iran does have a presidential election, but the list of candidates must be approved by the unelected and theocratic leadership of The Supreme Leader who is responsible for delineation and supervision of the policies of the Islamic

Type	Examples
Democracies	1. Germany, 2. Japan, 3. USA, 4. Japan, 5. India, 6. South Korea
False democracy	7. Russia,
Totalitarian	8. China,
Theocracy	9. Iran
Absolute monarchy	10. Saudi Arabia

Figure 7.6 Summary of country political arrangements

Table 7.4 Composition of country GDP

		Agricultural	Industrial	Service
		%	%	%
High income	USA	0.9	19.1	80.0
	UK	0.7	20.1	79.2
	France	1.7	19.5	78.8
	Germany	0.7	30.7	68.6
	Japan	1.1	30.1	68.7
	Canada	1.8	28.6	69.6
	South Korea	2.2	39.3	58.3
	Saudi Arabia	2.6	44.2	53.2
Middle Income	Indonesia	13.7	41.0	45.4
	Brazil	6.6	20.7	72.7
	China	7.9	40.5	51.6
	India	15.4	23.1	61.5
	Russia	4.7	32.4	63.4
	Iran	11.2	40.6	48.2
Low Income	Central African Republic	43.2	16.0	40.8
	Liberia	34.0	13.8	52.2
	Burma	24.1	35.6	40.3
	Ethiopia	34.8	21.6	43.6

Source: CIA. www.cia.gov/the-world-factbook/field/gdp-composition-by-sector-of-origin/.

Republic of Iran. The elected president has limited power compared to the Supreme Leader.

Country Economic Structures

The economic structures of the various countries in the world vary enormously in a number of ways. Key aspects of this include the following:

- The size of the economy
- The sectoral configuration of economies between agriculture, manufacture, and services

- The extent of government involvement in the management of the economy
- The balance between domestic consumption and export to other countries, and so on

Once again, these economic structures will have an influence on the extent to which a particular country is able/willing to implement its agreed climate change policies and the degree of success in collaborating with other countries on climate change issues.

Economic Sectors

In looking at economic structures within countries, a start point would be to look at the balance between the three main sectors of economies, namely, agriculture, manufacture, and service. This is illustrated in Table 7.4 showing the composition of country GDP for a sample of countries broken down into high-, medium-, and low-income countries. The countries shown in bold are those listed as being the 10 highest emitters of GHG.

This table shows a number of quite clear trends.

- In high-income countries, the agricultural component of a country's GDP is tiny while it is larger in middle-income countries and larger still in low-income countries.
- Low-income countries have much smaller industrial sectors compared to high- and middle-income countries.
- All of the countries which are in the top 10 highest emitters of greenhouse cases are either high income or medium income. None of them are low-income countries. This is because of the small industrial sector in these countries.

Economic Output

The size of a country's economy is generally measured by what is termed as gross domestic product (GDP). This measure reflects the size of the productive output of the economy covering all sectors of agriculture,

Table 7.5 Gross Domestic Product (GDP) of countries

Country	GDP 2021 (£bn)	Population 2022 (million)	GDP per head of population ($)
China	17,734	1439	12,324
USA	25,035	338	74,068
India	3469	1420	2443
Russia	2133	144	14,813
Japan	4301	124	34,685
Iran	360	88	4091
Canada	1990	38	52,368
Germany	4031	84	47,988
Saudi Arabia	1011	36	28,083
South Korea	1734	52	33,346

Source: Author constructed from several sources.

manufacture, and services. This is illustrated in Table 7.5 for the top 10 polluting countries.

Two points need to be mentioned here:

- Among the top 10 polluting countries, there will be huge variations in GDP per population and this will impact on levels of affiance and poverty.
- While India has the lowest GDP per capita from this group of countries, there are 57 other countries in the world with a lower GDP per capita than India.

GHG Emission Rates Per Capita

This has been included here because it indicates something about the economic structure of various countries. Table 7.6 shows the results for the top 10 polluting countries plus some low-income countries.

This table clearly shows the emissions per capita of the 10 largest polluting countries (including India) are times greater than those countries who have very low per capita emission rates. This is, of course, a consequence of the economic structures of the countries shown. The large polluting countries are all high- or middle-income

Table 7.6 GHG emissions per capita (2020)

Country	Emissions per capita (Mt)
China	8.2
USA	13.68
India	1.74
Russia	11.64
Japan	8.39
Iran	8.26
Canada	14.43
Germany	7.72
Saudi Arabia	16.96
South Korea	12.07
Central African Republic	0.07
Liberia	0.25
Ethiopia	0.15

Source: World Population Review. https://worldpopulationreview.com/country-rankings/carbon-footprint-by-country.

countries, which have "well-developed" economies resulting in large-scale consumption of energy, raw materials, and so on. The three African countries with very low pollution rates have very "under-developed" economies resulting in low levels of consumption. Certain words above are shown in parenthesis because it is a legitimate question to ask as to whether the economies of rich countries which are killing the planet should morally be called "well developed". Highly toxic might be a better label.

Country Inequalities

We have already seen in Chapter 3 that there are huge social and economic inequalities in the world. These inequalities may be between countries and within countries as will be discussed below.

Inequalities Between Countries

Inequalities of income and wealth are pronounced between various countries in the world. Since the 1990s, total global inequality was declined for the first time since the 1820s, but the impact of the Covid pandemic has reversed this trend.

In economic terms, inequalities can be looked at in terms of income and wealth. In summary, we see that

- **Income:** The richest 10 percent of the global population currently takes 52 percent of global income, whereas the poorest half of the population earns just 8 percent of it.
- **Wealth:** These are even more pronounced than income inequalities. The poorest half of the global population barely owns any wealth at all, possessing just 2 percent of the total. In contrast, the richest 10 percent of the global population own 76 percent of all wealth.

However, inequalities are not randomly scattered across the countries of the world. I have already noted above that countries can be categorized as high-, medium-, and low-income countries. While there will be variations in income and wealth between countries in a particular category, these are dwarfed by the variations between the categories themselves.

Inequalities Within Countries

One aspect of inequality concerns the distribution of income within a country. The Gini coefficient measures the extent to which the distribution of income within a country deviates from a perfectly equal distribution. A Gini coefficient of zero expresses perfect equality where everyone has the same income, while a coefficient of 100 expresses full inequality where only one person has all the income. In practice, the Gini coefficient for individual countries varies from 23.2 (Slovakia) to 63.0 (South Africa) thus meaning that Slovakia is the least unequal country in the world and South Africa the most unequal. Over the last two decades while income inequalities between countries have declined, income inequality has increased within most countries and income inequalities are likely to be higher in poor countries than rich countries.

Table 7.7 Country Gini coefficients

Country	Gini coefficient	The inequality decile where each country sits
China	38.2	4th
USA	41.5	5th
India	35.7	3rd
Russia	36.0	4th
Japan	32.9	3rd
Iran	43.6	6th
Canada	33.3	3rd
Germany	31.7	3rd
Saudi Arabia	54.1	8th
South Korea	31.4	3rd

If we look at our top 10 polluting countries, Table 7.7 shows the Gini coefficients for each country.

It can be seen, quite clearly, that inequality varies considerably between these countries, and this may have implications for their attitudes toward climate change.

Possible Country Responses to Climate Change

The 10 countries discussed in this chapter are the largest emitters of GHG and taken together about for almost 70 percent of global emissions. If these "top 10" countries fail to control and reduce their emissions, then the reality is that it is almost irrelevant what the other 190 or so countries achieve.

In this section, I try to set out what the responses of these countries might be based on what we know about them. I readily accept that much of this will be speculation and riddled with uncertainty, but such speculation can be useful as a pointer to considering what might happen.

The level of commitment to net zero

To start with, it is enlightening to get a picture of where these countries currently stand in relation to their policies on climate change. There are two aspects to this as follows: (1) the strength of the commitment and (2) the projected date of achievement.

1. The strength of commitment by a country to net zero can be considered by the way it has expressed its commitment in terms of
   - The target being set in law
   - The target set in a policy document
   - The target being set via a political pledge
   - No document submitted
2. The projected timing of the achievement of net zero which can be expressed as being
   - By 2050 or earlier
   - After 2050
   - No indication

Combining these two dimensions together (Figure 7.7), we can obtain the following overall picture for the 10 countries with the highest levels of GHG emissions.

Overall, this does not seem to be an encouraging position given that it involves the biggest emitters of GHG. One country has no commitments, and some others have weak commitments and are pledged to a date beyond 2050 for the achievement of net zero.

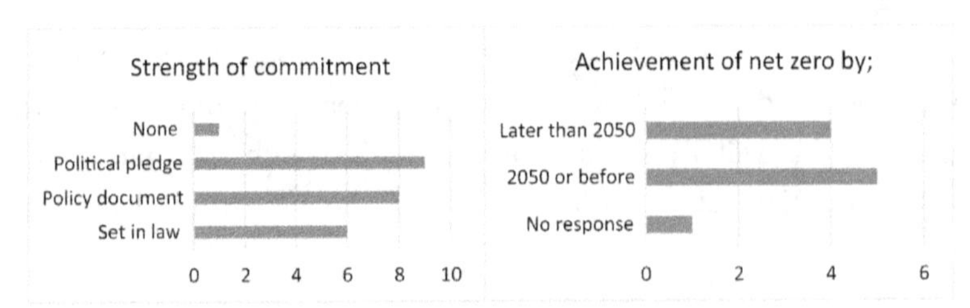

Figure 7.7 Commitment to and achievement of net zero

Source: https://eciu.net/netzerotracker.

Potential Costs to Countries

In general terms, I would suggest that each of these countries will need to consider the following "costs" they may encounter when deciding what they are really going to do about climate change as opposed to just making pledges, which may not be fulfilled. There are three aspects as follows:

1. **Financial costs:** The costs to the public purse of financing the actions required to mitigate climate change. Almost by definition, this means utilizing funds that might otherwise be used for public purposes such as health, education defense, and so on.
2. **Economic costs:** The impact on the country's economy of implementing the actions required to mitigate climate change. It is important that this doesn't just consider the impact on GDP but also the impact on GDP distribution. Some parts of the country and society may benefit from the implementation of climate change mitigation, but other parts may be seriously and negatively impacted.
3. **Political costs:** Apart from, perhaps, the most ruthless totalitarian state, it is the case that governments have to take some notice of public opinion in relation to their policies. To do otherwise may have a serious political cost.

Individual Country—Overall Situations

In Figure 7.8, I give a brief summary of the key issues which might influence a country's attitude to climate change for each of the top 10 polluting countries.

Collaboration and Trust Between Countries

One wonders about the lethargy of many countries in addressing climate change. In spite of the threat posed by climate change, is it that they are reluctant to make changes that will impact too strongly on

Country	Key issues
China	• Totalitarian state • Largest global polluter • Geopolitical ambitions which will require significant and ongoing military expenditure. • Declining and ageing population • Imperative to maintain or even increase economic growth rates
USA	• Largest polluter in terms of emissions per capita • Young and growing population which seems likely to pursue significant and ongoing economic growth. • Government is pursuing policies and large investments designed to achieve economic growth while mitigating climate change. • Democratic country but one where climate change is a political football
India	• Largest democracy in the world • Will soon have largest population in world but significant ageing. • Very high levels of poverty particularly in rural areas • Strong desire for economic growth • Has set a net zero target by 2070 but must be doubt as to achievability
Russia	• Fast becoming a totalitarian state • Largest land mass in the world but population is declining sharply. • Strong vested interest in the continued production of hydrocarbons • Existing policies indicate no real commitment to curb emissions. • Involved in prolonged war leading to international isolation
Japan	• A strong democracy • Strong targets for cutting emissions but will be challenging. • Island nature of Japan makes it very vulnerable to climate change. • Lack of renewable energy resources is problematic. • Ageing population will hamper economic growth
Iran	• Theocratic state with little respect for human rights • Regional ambitions require high levels of military expenditure. • Strong vested interest in the continued production of hydrocarbons • Little engagement with climate change issues
Canada	• Strong democracy • Major concerns about climate change especially British Columbia • Commitments to achieving net zero by 2050 but this may not be easy. • Strong vested interest in the continued production of hydrocarbons • Energy production and use in Canada is high but clean due to hydro power
Germany	• Strong democracy • Already been hit by climate change. • Pledged to achieve net zero emissions by 2045 and starting conditions for its achievement are excellent, • With strong political leadership and reasonable economic performance objectives should be achievable.
Saudi Arabia	• Absolute monarchy • Committed to achieving "net zero" GHG emissions by 2060. • Has numerous advantages in achieving this target? • Strong vested interest in the continued production of hydrocarbons
South Korea	• Strong democracy • Already experiencing the impacts of climate change • Declining and ageing population • High levels of GHG emissions relating to energy • Huge problem in getting away from fossil fuels

Figure 7.8 Collaboration and trust between countries

their own country without guarantees that other countries will do the same? Let us consider this question.

For a country, the costs associated with climate change mitigation will also need to be considered alongside the issue of the extent to which countries trust each other regarding climate change actions. This is best exemplified by what is termed the "the prisoner's dilemma" (Stanford University 1997) which is a well-known framework in game theory concerning how and why people cooperate or compete with each other. According to the prisoner's dilemma, rational individuals might not cooperate with each other although it would be in their combined best interests to do so. By prioritizing their personal interests, individuals acting rationally can create a worse overall result.

Climate change is a global problem, which requires actions by many people, companies, and countries around the world. As already noted, these actions will have cost impacts on the individual countries, whereas the benefits in terms of climate change mitigation are global and result from the collective actions of many. They will also affect future generations far more than today's generations.

Because the actions required are disruptive and require trust and collaboration between countries to work, some countries may choose to do only a limited amount of action and let the others take the strain. The problem of course is that if many or all countries take this view then there will be only limited progress on climate change mitigation. This prisoner's dilemma model might help us understand why countries across the world have postponed actions to limit further temperature rise for such a long time, despite scientists' warnings of the risks ahead.

Furthermore, if we think about collaboration and trust, the current picture doesn't look pretty. If we just consider our top 10 polluting countries, we, currently, see the existence of serious conflicts and disagreements between these nations as follows:

- **USA, Germany, and Canada versus Russia:** Relationships between these countries have deteriorated over several years but now can only be described as abysmal largely consequent on the hot war in Ukraine and the associated political and economic issues. The future of relations between these countries remains very unclear.
- **USA versus China:** This is an ongoing conflict which is currently just diplomatic and economic over the degree of economic and political influence across the world. This is

worsened by the threats of China to invade Taiwan, which would bring about an American military response.

- **India versus China:** There is an ongoing border dispute between India and China over disputed territory in the Himalayas. This has led to several military conflicts over the last 60 years; the last one being in 2022. Currently, the situation is stable, but both countries still have significant armed forces on the border.
- **Iran versus Saudi Arabia:** Relations between Iran and Saudi Arabia have been severely strained over several years because of geopolitical issues, such as aspirations for regional leadership, oil export policy and relations with the USA and other Western countries. In recent years, the two countries have been involved in a proxy war in Yemen where each country sponsors a different side in this civil war which has cost hundreds of thousands of lives. Recently, discussions brokered by China have attempted to improve relations.
- **Japan versus China:** Historically, relations were tense following the Korean War, the Cold War, and the grievances of Japanese war crimes, in World War II, committed in China and beyond. In recent years consequent on the expansion of trade, relationships have improved but there are still some geopolitical disagreements and tensions over Taiwan.
- **South Korea versus China:** There are two issues here. Firstly, South Korea is a strong ally of the USA but a proximate neighbor of China. Recently, China has practiced a pugnacious foreign policy that has caused tensions with its neighbors and the USA. Secondly, there is North Korea which is seen as a threat by South Korea. The landmine-strewn demilitarized zone is the world's most heavily fortified border and serves as a reminder that North Korea and South Korea are technically still at war. However, North Korea is strongly supported by China, which complicates its relationship with South Korea.

Given the above, it seems to me that, to a large extent, the relationships between many of these top 10 polluting countries are poor. This doesn't seem to be a good basis for collaborations on climate change.

Conclusions

In this chapter, I have examined the situation of the top 10 polluting countries which (as I keep mentioning) account for 70 percent of GHG emissions. I would suggest that the extensive and varied evidence I have presented in this chapter suggests that many or most of these countries will fail to achieve net zero by 2050. In the light of this and in spite of the urgent warnings from the IPCC which were mentioned at the start of the chapter, the likelihood of the world achieving net zero by 2050 looks to me to be very unlikely.

CHAPTER 8

What Really Needs to Happen?

Introduction

The conclusion of Chapter 6 was somewhat pessimistic about the chances of achieving net zero by 2050 and of containment of global temperatures. So, what really needs to happen?

Anyone studying the issues of climate change will soon come to realize that we are dealing with a massive complex issue in terms of the causes, the impacts, and the potential solutions. While this is undoubtedly true, it is not necessarily helpful. Another way to look at this is that getting away from the detail, it is actually quite simple to understand.

If we first consider the causes of climate change, then basically the problem is that over the last two to three centuries humanity has been accessing reserves of fossil fuels (coal, oil, and gas), which are found deep underground and have not been disturbed for millions of years. By accessing these fuels and using them either by burning (or some other way), humans have released huge volumes of main carbon-based GHG into the atmosphere, and these emissions have created the GHG effect which has resulted in rising global temperatures. Other activities of humanity such as mining, logging, and so on have also contributed to the related phenomena of biodiversity loss and habitat destruction. In this book, we have looked at some length at the impacts of global warming and climate change in relation to temperature rises, wild weather, shrinking of ice gaps, and many other weather phenomena.

When we turn to the solutions to global warming, we see a huge range of policies, projects, initiatives, and so on, which are aimed at reducing global emissions and resolving the climate change crisis. However, as I have already argued, I do not see these policies succeeding in achieving their declared objectives of net zero by 2050. If we fail to achieve this, the planet will continue to warm.

So, what should we be doing? In Chapter 1, I suggested that global warming can be considered in relation to three key issues:

- Population
- Consumption
- Capitalism and economic growth

In the sections below, I wish to consider how these three key issues might be addressed under the headings of

- Population control
- Lower resource consumption
- Alternative economic models

Population Control

Population Trends

Basically, the more people there are on Earth, the greater will be the need for the basic commodities of food, energy, housing, and so on. As has already been noted, the global population has mushroomed enormously in recent centuries rising from one billion in the 1804 AD to 8 billion today. It cannot be a coincidence that the rise in global temperatures has taken place during a period of major economic development and population growth.

Moreover, the population will still continue to rise for some time to come. The United Nations projects that the size of the world's population is almost certain to rise over the next several decades, but there is a degree of uncertainty associated with its projections. Later in the century, there is about 50 percent chance that the world's population will peak (that its size will stabilize or begin to decrease) before 2100. Subsequently, there is a probability of 95 percent that the size of the global population will lie between 9.4 and 10.0 billion in 2050 and between 8.9 and 12.4 billion in 2100. The key unknown here is the fertility rate of females in the future. With this population growth, will come a growth in the demands for a variety of commodities.

Carrying Capacity

What is needed here is an understanding of what is referred to as the Earth's "carrying capacity." Clearly, the Earth cannot support an infinite number of people; so what must be the limit? Have we already breached that limit as many think we have?

The problem is that the carrying capacity is linked to both the numbers of people on Earth and their level of consumption. In all probability, the Earth could support many more than the current 8 billion people as long as their lifestyles were very primitive, and they consumed low levels of the Earth's resources. Equally, it would not be possible for the Earth to support the current levels of population if everyone on Earth had the consumption levels of the most economically developed nations. As noted in Chapter 1, billions of people already live in poverty or extreme poverty and it doesn't need a great deal of imagination to see that, given the large numbers of humans involved, any improvements in alleviating this poverty would result in an enormous rise in demand for the Earth's resources. This is not an argument for ignoring the needs of the poor but an argument for redistribution of resources.

So, what is the carrying capacity of the Earth. Debates about this date back hundreds of years. The range of estimates is enormous, vaying from 2 billion people to more than one trillion people. Scientists disagree not only on the final number but more importantly about the best and most accurate way of determining that number—hence the huge variability. The differences in estimates come down, largely, to the issue of consumption already discussed. The problem, of course, is that to make such estimates, you have to make assumptions about consumption habits relating to energy, food, water, consumables, and so on. However, the majority of studies estimate that the Earth's carrying capacity is at or beneath 8 billion people (UNEP), which is well below the current and future projected population of the Earth.

Actions Needed

In April 2019, the Commission on Population and Development held its session in New York. According to this commission, population growth could still be controlled, and the situation could be reversed, and a balance could be achieved between population and available resources. A number of measures have been suggested including the following (Tomorrow City 2021):

1. **Empowering women:** Gender equality should be a central theme of social, economic, or environmental interventions carried out by governments, companies, and civil society. This is the only way to allow women to manage their lives in an autonomous manner and to decide for themselves whether they want to continue studying, enter the labor market, form a family, or have children.
2. **Universal access to quality education:** Controlling the population involves enjoying a good education. It is a fact that keeping girls in school makes them less likely to have children at a young age and, therefore, reduces the number of maternal deaths.
3. **Family planning programs:** Good sex education programs and the implementation of family planning programs that enable access to contraception reduces the number of unwanted children and, therefore, improves the lives of members of the family unit, allowing them to optimize their resources and avoid situations of poverty, hunger, or disease.
4. **Implementing more efficient production methods:** Investing in research, development, and innovation to develop production methods that improve the efficiency of human activities, will enable the amount of natural resources required to meet the needs of the population to be reduced.
5. **Controlling migration flows:** Establishing programs that prevent indigenous people from having to leave their homes for work reasons, or as a means of survival or due to armed conflicts, enables a more sustainable use of resources and prevents resource depletion. It also improves the lives of communities from the moment it prevents families from being separated and reduces alienation and depopulation.

Now, these are five key actions and there may be others. However, it is quite clear that those countries where population growth is highest are the least well equipped to implement these actions. This would require substantial investment from developed countries. The chances of this happening to any great degree seem low.

Lower Resource Consumption

The world has enough for everyone's need, but not enough for everyone's greed.

—Mahatma Gandhi

The demands from individuals in the affluent world for commodities go well beyond basic needs for survival. Modern life means that individuals, particularly in developed countries, have the opportunity to acquire a wide range of goods, services, and energy. This has been the case in the West for decades but now we see the rise in consumerism in other parts of the world such as China and India, with large and increasingly affluent populations living alongside great poverty.

Production of these goods and services often requires high volumes of raw materials, energy, and so on. Furthermore, it should be noted that, in some cases, overconsumption of certain commodities in affluent countries can have negative consequences.

Hence, reducing consumption is a key action in addressing climate change and the associated phenomena.

When discussing this, the term 3Rs often appears which stands for reduce, reuse, and recycle. I discuss this below, but I would like to add a fourth "R" of repair and also change the order in which these are considered.

Reuse

This can occur in a number of ways. For example, people might reuse the same items or materials several times such as reusing food containers and bottles instead of throwing them out or using travel coffee mugs instead of single-use cups. The motivation for doing this is that there is often a small financial penalty for not reusing an item. A second approach might be where people buy second-hand goods from a previous owner which means an item is being reused by a different person. The stigma of buying second hand appears to be lifting, and there are several websites facilitating such transactions. Finally, an item may be used for a different purpose such as using a saucepan as a flowerpot.

Reports from organizations supporting reuse suggested that the level of reuse activity started to decline some years ago (Vaughan 2014), and this situation worsened during the Covid pandemic. However, there seems to be huge scope for a much higher level of reuse in our society undertaken in conjunction with repair which is discussed next. Businesses can assist here with proper pricing. This is an anecdote, but I use a printer which costs new (including two printer cartridges) less than the cost of replacement cartridges themselves.

Repair

You know the scenario. Your tumble dryer stops working and the repair operative tells you it can't be repaired as there are no parts available or the cost of repair (labor and parts) is only a little less than the cost of a new machine. You only have to visit a recycling site to see the number of domestic appliances or electronic goods which have been disposed of, although many of them could probably be easily repaired. In the UK, a new 'right to repair' law aimed to tackle the scale of the problem by making household washing machines, washer dryers, dishwashers, fridges, TVs, and other electronic displays easier to repair. Manufacturers are now required to make spare parts available for appliance repairs for a minimum of 7 to 10 years after an appliance stops being manufactured. It remains to be seen what impact this has but it is a step in the right direction.

Recycle

Waste materials can be collected and separated out into their different component ingredients, and these can then be remade into different products. The basic phases in recycling are the collection of waste materials, their processing or manufacture into new products, and the sale of those products, which may then themselves be recycled. Typical materials that are recycled include iron and steel scrap, aluminum cans, glass bottles, paper, wood, and plastics.

Recycling can be thought of as internal and external.

- **Internal:** This is the reuse in a manufacturing process of materials that are a waste product of that process. Internal recycling is common in the metals industry, for example.
- **External:** The reclaiming of materials from a product that has been worn out or rendered obsolete. An example of external recycling is the collection of old newspapers and magazines for repulping and their manufacture into new paper products. Aluminum cans and glass bottles are other examples of everyday objects that are externally recycled on a wide scale.

Energy is required to recycle the product and to change its physical properties into something totally different. For example, the plastic from bottles might be made into public benches or fleece jackets. Unfortunately, these costs sometimes mean that it is cheaper to buy a new product than recycle the old.

There has been a lot of publicity and emphasis on recycling with lots of community groups springing up to take actions. While this is laudable, its importance should not be overstated. One study (Crunden 2021) analyzed more than 80 separate means that could help keep the world from passing the oft-cited threshold of 1.5°C or 2.0°C of warming and concluded that the recycling industry's projected contributions fell below the median, trailing geothermal power, efficient aviation, forest protection, and dozens of other actions.

Reduce

This is discussed last because of an observation that while there are many initiatives to be seen concerning reuse, repair, and recycle, and many people who seem committed to these three actions, I am not sure, the same attention is given to reduce. A recent report stated that while the US population is 60 percent larger than it was in 1970, its consumer spending is up 400 percent (adjusted for inflation) and other rich nations aren't much better.

Let's look at some recent statistics starting with food waste.

- The UK throws away 9.52 million tons of food per year.
- This waste emits 25 million tons of CO_2 more than Kenya's total annual emissions.
- 40 percent of food produced globally goes to waste while billions of people go hungry.

At the same time, large proportions of the populations in developed countries around the world are overweight or obese leading to all sorts of health problems.

Now, let's look at transport. For the last 20 years, road mileage driven in all vehicles has increased annually. In 2021, 297.6 billion vehicle miles (bvm) were driven on Great Britain's roads by all vehicles, a 11.9 percent increase on the previous year. However, traffic in 2021 was 12.1 percent lower compared to 2019 prepandemic levels, and it is unclear what future trends will be. Just focusing on cars, the mileage driven on our roads equates to almost 5000 miles for every man, woman, and child. Now much of this mileage is, effectively, unavoidable for a variety of reasons, but we should be asking ourselves whether other journeys are really necessary.

If we look at the situation with aviation, we see that between 2001 and 2018, flights from the UK and overseas increased by 52 percent for holidays and for trips to visit friends. Business flights increased by only 10 percent. The numbers declined drastically during Covid but have started to recover. A similar situation will be found in many other developed countries especially the USA where there are very high levels of internal flights. The question we have to ask is are all these trips really necessary using the most polluting form of transport?

Another area to consider is the curse of fashion clothes for men, women, and children. Now it could be argued that the main purpose of clothing is to keep us warm and protect us from the elements. On that basis, we could manage with fairly few items of clothing which would last for several years. However, in reality, we all own far more items of clothing than we really need, and we are encouraged to be fashionable and purchase new and fashionable items of clothing each and every year. This is clothing we don't need but we want.

Of particular concern is the idea of fast fashion which can be defined as the quick turnover of fashion trends and the move toward cheap, mass-produced clothing with new lines constantly released. The problem is that fast fashion is toxic to the environment, and the UN suggests that overall, the fashion industry is responsible for 8–10 percent of global GHG emissions, which is more than aviation and shipping combined. Also, the World Bank estimates that global clothes sales could increase by up to 65 percent by 2030.

Most of fashion's environmental impact come from the use of raw materials. A few examples are.

- Cotton for the fashion industry uses about 2.5 percent of the world's farmland.
- Synthetic materials like polyester require an estimated 342 million barrels of oil every year.
- Clothes production processes such as dying require 43 million tons of chemicals a year.
- The industry also uses a lot of water. The UN estimates that a single pair of jeans requires a two pounds of cotton and because cotton tends to be grown in dry environments, producing this kilo requires about 1700 to 2200 gallons of water. That's about 10 years' worth of drinking water for one person.

It is clear that a big reduction in the amount of fashion clothes purchased could have a significant impact on global emissions.

Finally, there are many other areas where we can reduce our consumption such as switching our house thermostats to a lower temperature setting or making less use of air-conditioning systems.

Now there are a couple of complicating factors about this that we need to keep in mind. Firstly, on a personal note, the author would emphasize the importance of "reduce" over the other 3Rs, but this is difficult because people don't want to give up their material goods purchases. Secondly, as I will discuss below, the capitalist model of the economy, which is so common throughout the world, requires ongoing economic growth for stability, which in turn requires people to spend more by acquiring more goods and services. Hence, we frequently see governments exhorting or incentivizing people to go out and spend more to "help the economy." We see concerns about the reduction in the numbers of new cars sold because that will hurt the economy. We see encouragement for people to visit restaurants in order to support the restaurant economy. During the pandemic, the then Chancellor, Rishi Sunak, started a scheme encouraging people to go out and eat in a local restaurant by providing a financial subsidy. At the end of pandemic, there were rallying cries to boost the economy by opening our wallets. Shopping has been cast as a positive act, retail therapy a civic duty. We cannot escape the fact that major disruptions in consumption will severely dislocate the economies in many countries, which would have huge political implications. However, there might be another way as discussed by MacKinnon (2021) in his book *The Day the World Stops Shopping*.

Secondly, while people in developed countries are overeating, wasting food, buying excessive amounts of clothes, and flying around the world, billions of people live in abject poverty. While there are exceptions to this rule, it seems to me that the majority of people in developed countries react to this by either ignoring it completely or saying, "isn't it terrible" and making a contribution to a charity or encouraging the government to give more overseas aid. Of course, the same situation arises in many poor countries where the rich in the countries take the same attitude to their poor as people in developed countries. The problem here is that while in the past, we were able to keep buying, spending, and consuming a huge variety of items without worrying too much about the poor this is no longer the case. The "chickens have come home to roost" and our consumption habits are now damaging the planet and threatening humanity.

This leads me on to my third theme of alternative economic models.

Alternative Economic Models

Capitalism has been the dominant economic model in the developed world for several centuries. It can take a variety of formats and has subsequently spread to most parts of the world. Over the years, some countries, most notably USSR and China, have tried to replace capitalism with other economic models, but this has largely been a failure and they have reverted to their own versions of capitalism.

The essential feature of capitalism is the motive to make a profit. In a capitalist economy, assets, such as factories, mines, and railroads, can be privately owned and controlled, labor is purchased for money wages with capital gains accruing to private owners.

Now not everyone will agree with this but there is a strong argument for saying that capitalism is the most successful economic system ever developed by humankind. In the developed world, lots of people forget that we now live in times of unprecedented prosperity and, even in developing countries, people are better off than they were 50 years ago, albeit they are still poor. Before capitalism emerged, most people were trapped in extreme poverty. Most remarkably, in recent decades, the decline in poverty has accelerated at a pace unmatched in any previous period of human history. In the UK prior to the Industrial Revolution, most people lived on about £2.20 per day or £800 per year (in today's money) while in 2015, median average earnings for a fulltime UK employee were £32,500. However, there is, of course, a downside to capitalism and themes that come to mind include exploitation of the labor force, insecurity of employment, and extreme inequalities in income and wealth.

Why Do We Need Alternative Economic Models?

Whatever the merits and demerits of capitalism, be they moral or practical, there is now a bigger and overwhelming issue. The climate change crisis forces us to ask the question whether we can continue with such a system.

The reality is that in most years, the economies of most countries have grown in size (at differential rates) each year with occasional periods of economic recession resulting in a contraction in the size of the economy. As will be discussed later Chapter 10, the focus of most governments around the world is to achieve ongoing economic growth (the greater the better) in order to finance increasing standards of living for their populations. Such growth improves living standards and also generates funds for the expansion of public services. However, this ongoing economic growth has serious implications for the climate.

While capitalism has been the major economic model for several centuries, it might be that climate change and the need to reduce global warming mean that it is no longer "fit for purpose." Hence, we must be prepared to ask the difficult question as to whether traditional capitalism can survive the problems posed by climate change. This is a big issue and can only be summarized here. Should we be looking for alternative economic models?

Now the mention of alternative economic models will usually bring about a furore of objections and insults from a wide range of people including right wing politicians, bankers, billionaires, and business chief executives. People who propose such changes will be called a variety of things including lunatics, deluded, communists, Marxists, traitors, and so on, and this is just for raising the issue for discussion. Interestingly, many of the people who will say these things are often the same people who promote climate change denial as outlined in Chapter 5. One wonders if vested interests are at play here!

Shared Socioeconomic Pathways

In the late 2000s, researchers from different modeling groups around the world began the process of developing new scenarios to explore how the world might change over the rest of the 21st century. As mentioned in Chapter 3, they developed a series of Shared Socioeconomic Pathways (SSPs) scenarios of projected socioeconomic global changes up to 2100. The scenarios are as follows:

- SSP1: Sustainability (Taking the Green Road)
- SSP2: Middle of the Road

- SSP3: Regional Rivalry (A Rocky Road)
- SSP4: Inequality (A Road divided)
- SSP5: Fossil-Fueled Development (Taking the Highway)

These scenarios examine how global society, demographics, and economics might change over the next century and provide narratives describing alternative types of developments for each pathway. The pathways are summarized below.

- **SSP1: Sustainability (Taking the Green Road)**: The world shifts gradually, but pervasively, toward a more sustainable path, emphasizing more inclusive development that respects perceived environmental boundaries. Management of the global commons slowly improves; educational and health investments accelerate the demographic transition, and the emphasis on economic growth shifts toward a broader emphasis on human well-being. Driven by an increasing commitment to achieving development goals, inequality is reduced both across and within countries. Consumption is oriented toward low material growth and lower resource and energy intensity.
- **SSP2: Middle of the Road:** The world follows a path in which social, economic, and technological trends do not shift markedly from historical patterns. Development and income growth proceeds unevenly, with some countries making relatively good progress while others fall short of expectations. Environmental systems experience degradation, although there are some improvements, and overall, the intensity of resource and energy use declines. Income inequality persists or improves only slowly and challenges to reducing vulnerability to societal and environmental changes remain.
- **SSP3: Regional Rivalry (A Rocky Road)**: A resurgent nationalism, concerns about competitiveness and security, and regional conflicts push countries to increasingly focus on domestic or, at most, regional issues. Policies shift over time to become increasingly oriented toward national and regional security issues. Countries focus on achieving energy and

food security goals within their own regions at the expense of broader-based development. Investments in education and technological development decline. Economic development is slow, consumption is material-intensive, and inequalities persist or worsen over time. Population growth is low in industrialized and high in developing countries. A low international priority for addressing environmental concerns leads to strong environmental degradation in some regions.

- **SSP4: Inequality (A Road Divided)**: Highly unequal investments in human capital, combined with increasing disparities in economic opportunity and political power, lead to increasing inequalities and stratification both across and within countries. Over time, a gap widens between an internationally connected society that contributes to knowledge- and capital-intensive sectors of the global economy, and a fragmented collection of lower income, poorly educated societies that work in a labor-intensive, low-tech economy. Social cohesion degrades and conflict and unrest become increasingly common. Technology development is high in the high-tech economy and sectors. The globally connected energy sector diversifies, with investments in not only carbon-intensive fuels like coal and unconventional oil but also low-carbon energy sources. Environmental policies focus on local issues around middle- and high-income areas.
- **SSP5: Fossil-Fueled Development (Taking the Highway)**: This world places increasing faith in competitive markets, innovation, and participatory societies to produce rapid technological progress and development of human capital as the path to sustainable development. Global markets are increasingly integrated. There are also strong investments in health, education, and institutions to enhance human and social capital. At the same time, the push for economic and social development is coupled with the exploitation of abundant fossil fuel resources and the adoption of resource and energy intensive lifestyles around the world. All these factors lead to rapid growth of the global economy, while global population peaks and declines

in the 21st century. Local environmental problems like air pollution are successfully managed. There is faith in the ability to effectively manage social and ecological systems, including geo-engineering if necessary.

Referring back to Chapter 3, the first thing to note is that it is only SSP1, the very low emissions scenario, which keeps global temperature rises close to 1.5°C and below 2.0°C. As noted in Table 3.1, the other scenarios show increasing divergence of global temperatures away from 1.5°C/2.0°C. Also, SSP1 is the only scenario which might move away from the traditional capitalist model with a shift in emphasis from headline economic growth toward a broader emphasis on human well-being. Many economists have argued for many years about the error of using economic growth as the headline figure for economic performance instead of considering improvements in human well-being. For example, an empirically researched paper published by Howarth and Kennedy (2026) suggests that, in recent decades, substantial growth in per capita income has failed to generate commensurate increases in social welfare. Much similar evidence on this matter can be produced but the impacts on politicians and political discourse remains unaltered.

In light of this, we can consider just a few alternative economic models, which have been posed.

What Alternative Economic Models Are There

Based on earlier discussion, it would be generally agreed that capitalism has three key principles.

- The majority of "the means of production" (land, resources, capital) are concentrated in private hands.
- The majority of this work for a wage (i.e., for other people).
- Markets are used to mediate between producer and consumer (set prices and so on).

Although there is agreement that the alternative economy also refers to processes of production, exchange, and consumption, they differ markedly from the so-called capitalist mainstream in a number of regards.

There are quite a number of alternative economic models, which have been developed, and some of these are summarized, briefly, below being.

- Economic democracy
- Common good economy
- Doughnut economics
- Post growth economy

Economic Democracy

This approach places the control of enterprises and the means of production, placing resources, factories, and other productive capital into the hands of the people and away from the short-term interests of both the state and private sectors. This form of economics would not need to rely on growth.

Under the vision of economic democracy, workers would control most enterprises democratically. To change this form of enterprise structure, legislation and subsidies could support us to buy the companies we work in through labor trusts and leveraged buyouts, coupon-based markets or "share levies" on corporate profits.

Bankrupt companies would be restructured as worker self-managed. Enterprises like Spain's Mondragon Co-operative Group have been shown to be more efficient than most private companies. Statewide examples of co-operative economics include the Quebec Social Economy. Workers would control (but not own) the enterprises they work in and, after paying a "capital assets tax" (a sort of rent) on revenue-generating property, any surplus would be divided democratically between them.

Enterprises would still interact with one another and with consumers in a market driven by the forces of supply and demand. Innovation and entrepreneurial activity would flourish. This economic model would not tend to the need for hyper-consumerism and industry could be guided by market frameworks that shift enterprises from seeing products themselves as benefits to seeing production as a cost of delivering to real (not created) well-being needs.

Common Good Economy

This, as the name already suggests, is based on a constitutional merit of common welfare/common good, which can be found in most constitutions around the world. At its core, it advocates that the economy is there for the common good, not a privileged elite, which, arguably, contradicts our current economic system.

The Common Good Matrix is an instrument that helps projects and organizations to identify and show their contribution to society. The matrix reports and evaluates the impact of the project on people, environment, and society. It shows the results in a clear, transparent, and comparable way. An Economy for the Common Good has the following features:

- Growth is no longer necessary, and organizations are liberated from "eating or being eaten" but aims for an optimal size.
- Cooperation is rewarded.
- Meaningful jobs are created.
- Money is only a means and not an end in itself.
- An economic model based on values.
- An economy that does not start from scarcity but strives to create abundance for everyone.

The matrix includes five main categories: human dignity, solidarity and justice, ecological sustainability, transparency, and participation. The matrix is based on a point-receiving system and can be applied to communities, companies, schools, and even private households. The goal is to reward institutions with a high positive score with benefits like tax reduction, while those institutions that score below an average can get a tax increase.

Doughnut Economics

This is a holistic economic system conceived by the English economist Kate Raworth. The principal idea concerns humanity's biggest challenge being to meet the needs of all within the means of the planet. This means ensuring that no one falls short on life's essentials (from food and housing to healthcare and political voice), while ensuring that collectively we do not overshoot our pressure on Earth's life-supporting systems, on which we fundamentally depend such as a stable climate, fertile soils, and a protective ozone layer.

The Doughnut of social and planetary boundaries is a chart that visualizes the impact of our current global economic system. It depicts nine planetary boundaries which set out the constraints our actions and twelve dimensions of what are termed a social foundation meaning basic needs. This is set out in Figure 8.1.

The thrust of doughnut economics is to change the goal from endless GDP growth to meeting basic needs for all. At the same time, see the big picture by recognizing that the economy is embedded within, and dependent upon, society and the living world. Doughnut Economics recognizes that human behavior can be nurtured to be cooperative and caring, just as it can be competitive and individualistic.

Post Growth Economy

Postgrowth is a worldview that sees society operating better without the demands of constant economic growth. It proposes that widespread economic justice, social well-being, and ecological regeneration are only possible when money inherently circulates through our economy. This deals with the central problem of capitalism in an environment of climate change that it requires growth to be stable. It centers around the idea of reducing consumerism, supporting repairing skills and promoting regional economic production. In short, PGE is based on principles of deceleration, decluttering/moderation, communal use/sharing production-consumption, repairing skills, and a fruitful local economy. All of these lead to an increase in social justice on a global scale while remaining within the ecological boundaries. This is at its core, the

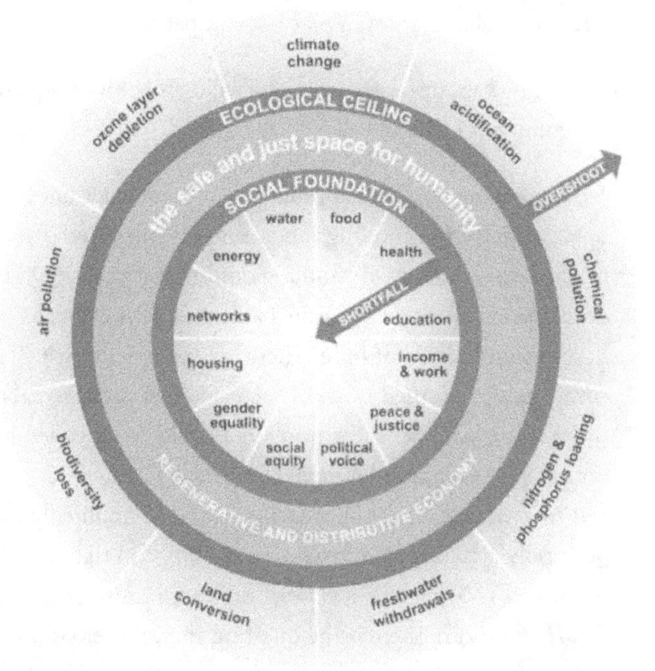

Figure 8.1 Doughnut economics

purest form of a sustainable (environmentally and humane) economy. The goals of the PGE are to reduce/bisect and reorganize the current industrial production.

Conclusion on Alternative Economic Models

We have discussed, briefly, the basics of four alternative economic models and there are quite a few more. If we attempted to synthesize some key themes which are part of some or all of these alternative models, we might see the following themes:

- Not focused on economic growth.
- Stronger focus on human/community well-being.
- Common or community ownership of productive resources as opposed to private ownership by individuals.

- Focus on collaboration in place of competition.

Now reading about these models will probably lead many to conclude a number of things.

- ***Underdeveloped:*** These approaches are somewhat underdeveloped, undertested and might be seen as being naïve. There might be a lack of clarity and big doubts that they could be developed to work on a national and international scale, at least not, in short period of time. There are examples of where a small company has adopted some of the principles discussed above but the doubt is whether these principles could be extended to larger domestic or even international countries.
- ***Political attack:*** These approaches would undoubtedly be politically attacked from parties of the political right and be rubbished as being too "socialist" or "communist."
- ***Public finances:*** There is the question that, if there is no economic growth, how do we generate tax revenues to finance improvements in health and social care to deal with an ageing population.
- ***Public support:*** These approaches will probably struggle to get enough support from voters in a country who just don't want this form of society which might stop them getting richer by their own efforts.

Conclusion

In this chapter, I have tried to focus on three key themes which go well beyond current international polices on climate change but are fundamental to solving the problem. These themes concerned population control, lower resource consumption, and alternative economic models. I suspect all three of these themes would be politically controversial and would be highly unpopular in many quarters.

Theme 1 on population control requires large-scale investment, which could only be provided by high-income countries. However, we have seen from the COP summits that even when funds are pledged for climate change mitigation and adaptation, the money doesn't always appear. There are similar problems in health. The Covid pandemic has raised concerns about the robustness of low-income countries in preventing the transmission of viruses within their countries and internationally. Again, international investment is needed but unlikely to happen. This is for the simple reason that high-income countries will need all of the resources they can muster to deal with climate change in their own country.

Regarding theme 2 on promoting lower resource consumption, there are a wide variety of actions taking place in many countries in relation to the 4Rs and these are hugely publicized in the media. Let us not knock these actions for they are worthwhile and need to be encouraged. However, what is currently happening in relation to reduced food consumption, reduced energy consumption, reduced fashion, and so on is a drop in the ocean compared to what is really needed. As I said in the introduction, it seems to me that of people are being led to believe that if they switch to an electric car, recycle their plastic bottles, and eat a bit more vegan food, then that will be sufficient. Clearly, the need for reduced consumption goes way beyond this and means major changes in lifestyles which would be extremely unpopular and probably impossible to implement in democratic countries.

Regarding theme 3 on alternative economic models, my perspectives are that while, in the context of climate change, there is a pressing need for an economic system which is not predicated on ongoing economic growth and consumption of the Earth's resources, I just cannot see much progress being made in the near future. The reality is that we are dealing with massive, vested interests in the existing capitalist model both economic and political and I cannot see how these can be overcome. Rightly or wrongly, I think these new models will be associated with the failed planning approaches of Russian and Chinese communism and dismissed accordingly. I think my conclusion must be that there is zero chance of any of this happening, particularly in high-income countries, for the simple reason that it is politically impossible to do so. Unfortunately, this still leaves us with the problem of a growth-driven capitalist system in an era of GHG-driven climate change.

Many will argue that the solution lies in "reforming" capitalism by means of future technological developments. However, I have argued several times in this book that this is a dangerous course of action. Some of these technological developments may never come to fruition or may become available but a long time in the future by which time the damage to the Earth may have been done. Also, even with such technological developments, the capitalist model may still function using the Earths nonrenewable resources (e.g., rare Earth elements) to the point of exhaustion. Some will argue that such resources may then be acquired from mining the moon, a suggestion which I see as fanciful given the economics of space flight which still involves firing people into space in a vehicle sitting on top of half a million gallons of highly flammable liquid gases.

CHAPTER 9

Future Scenarios

Introduction

In Chapter 7, my analysis of key countries and their likely attitudes toward climate change suggested that it was very unlikely that the world would achieve net zero by 2050. In Chapter 8, I outlined what I saw as the key measures that need to be applied in order to mitigate climate change, namely, population control, lower resource consumption, and alternative economic models. I expressed strong doubt that any of these would be seriously implemented.

In this chapter, I now try to outline some future scenarios based on what I see as the following:

- Likely impacts on planet Earth if we fail to contain temperature rises to within 1.5°C or 2.0°C
- Possible responses to the situations described above
- Conclusions

Likely Impacts

Some of these have already been discussed but are mentioned again for the simple reason that while they have already happened, to some extent, they may get worse. Others haven't yet happened and may be seen as speculative.

Extreme Weather

The term extreme weather or wild weather is used to describe particular weather events which are significantly different from the average weather pattern for an area. There are two main categories of extreme weather.

- ***Weather-related:*** These are various but shorter incidents like tornadoes, deep freezes, or heat waves.
- ***Climate-related:*** These last longer or are caused by a build-up of weather-related events over time.

Extreme weather events can be caused by natural cycles or more immediate influences like high pressure systems. However, they are becoming more likely because of the climate crisis caused by the human emission of GHGs.

Various types of extreme weather events that can have a serious impact on human populations and which are impacted by climate change. Climate change itself is making many extreme weather events both more frequent and more severe. A new scientific field called "extreme event attribution" has emerged to assess the human fingerprints on any particular extreme weather event, such as a storm or heat wave. Scientific evidence (Carbon Brief 2022) found that 70 percent of these events were made more likely or more severe by human-caused climate change. There are several reasons that climate change has this effect, and they vary depending on the weather event in question.

Aspects of extreme weather include the following:

- **Drought:** A drought is defined as "a period of abnormally dry weather sufficiently prolonged for the lack of water to cause serious hydrologic imbalance in the affected area." Such a spell lasts long enough to diminish the water supply or damage crops. Climate change raises the likelihood of drought because higher temperatures lead to more evaporation, reducing surface water and drying out soils and vegetation. Furthermore, increased winter temperatures means that less precipitation falls as snow in some areas.
- **Heat waves:** A heat wave is a stretch of unusually hot weather that lasts for two or more days. To be considered a heat wave, temperatures must rise above the average for an area, so a normal day in Madrid might be regarded as a heatwave is Helsinki. As global temperatures rise, periods of extreme heat naturally

increase. In recent years, record high temperatures have occurred in many countries.

- **Winter storms:** Winter storms are a life-threatening combination of heavy snow, blowing snow, and dangerous wind chill. A blizzard is a type of winter storm that combines heavy wind with blowing snow to severely reduce visibility. While climate change may lead to warmer winters overall, it can still increase the amount of snow that falls during winter storms. This is because it increases the amount of moisture in the atmosphere, which will fall as snow if temperatures dip.
- **Hurricanes (including typhoons and cyclones):** These are a type of storm that forms over tropical or subtropical waters. They consist of a rotating circle of clouds and thunderstorms. Hurricanes can be classified by their wind speed with a Category 1 being between 75 and 95 mph while a Category 4 being between 130 and 156 mph. Climate change has had several impacts on tropical storms. Warmer sea temperatures make storms both stronger and wetter, and it is predicted that the number of the strongest Category 4 and 5 hurricanes will increase as the climate warms. How climate change will impact the overall frequency of hurricanes is less clear.
- **Wildfires:** These are unplanned, uncontrolled fires that burn in a natural area like a forest or grassland. They can start either because of a natural occurrence like a lightning strike or because of human activity. However, their spread is determined by external conditions such as high temperatures, high wind speeds, and low precipitation that leave lots of dried vegetation as fuel. Wildfire risk is increasing for much the same reason as drought. Climate change raises temperatures, increases the amount of dry vegetation that serves as wildfire fuel, and reduces the amount of water available.

Rising Sea Levels

Sea level rise is caused primarily by two factors related to global warming: the added water from melting ice sheets and glaciers, and the expansion of seawater as it warms. Even under the most modest scenarios, climate change scientists expect sea level rises to affect the Earth's coastlines significantly. A growing knowledge base is why scientific organizations like the IPCC are publishing sea level rise projections with increasing levels of confidence. In its 2019 report, the IPCC projected 0.6 to 1.1 meters of global sea level rise by 2100 if GHG emissions remain at high rates (NRDC 2019). If countries do cut their emissions significantly, the IPCC expects 0.3 to 0.6 meters of sea level rise by 2100.

It is difficult to predict exactly which coastal areas will be affected by these sea level rises but some projections can be made. By 2100, there will be severe difficulties in some developed countries, but the situation will be even worse in other parts of the globe such as Bangladesh and the pacific islands. The impact of storm surges linked to wild weather hurricanes could also exacerbate this situation.

Desertification

Desertification has been described as "the greatest environmental challenge of our time" and climate change is making it worse. It is an issue that reaches far beyond those living in and around the world's deserts, threatening food security, and livelihoods of more than two billion people. While the term desertification may bring to mind the windswept sand dunes of the Sahara or the vast salt pans of the Kalahari, it is more than that. The UN has defined desertification "as land degradation in arid, semi-arid and dry sub-humid areas resulting from various factors including climatic variations and human activities."

The combined impact of climate change, land mismanagement, and unsustainable freshwater use has seen the world's water-scarce regions increasingly degraded. This leaves their soils less able to support crops, livestock, and wildlife.

Looking first at the role of the climate, a significant factor is that the land surface is warming more quickly than the Earth's surface as a whole. So, while global average temperatures are around 1.1°C warmer now than in pre-industrial times, the land surface has warmed by approximately 1.7°C.

Climate change can affect the frequency and magnitude of extreme events like droughts and floods. In areas that are naturally dry, for example, a drought can have a huge impact on vegetation cover and productivity, particularly if that land is being used by high numbers of livestock. "As plants die off due to lack of water, the soil becomes bare and is more easily eroded by wind and by water when the rains do eventually come."

As well as physical impacts on the landscape, climate change can impact on humans because it reduces options for adaptation and livelihoods and can drive people to overexploit the land. That overexploitation refers to the way that humans can mismanage land and causes it to degrade. Perhaps, the most obvious way is through deforestation. Removing trees can upset the balance of nutrients in the soil and take away the roots that helps bind the soil together, leaving it at risk of being eroded and washed or blown away.

In general, the global area of drylands is expected to expand as the climate warms. Projections under moderate and high emission scenarios suggest that drylands will increase by 11 percent and 23 percent, respectively, compared to 1961–90 (Climate Change Post 2016). This would mean drylands could make up either 50 percent or 56 percent, respectively, of the Earth's land surface by the end of this century, up from around 38 percent today.

Loss of Species

Chapter 5 talked about the topics of biodiversity loss, habitat destruction, and natural resource depletion, which are inextricably linked to climate change. It has already been noted that global warming has the potential to cause extinctions in many of the world's especially valuable ecosystems. Depending how a species responds to global warming, especially their ability to migrate to new sites, habitat change in many eco-regions has the potential to result in catastrophic species loss. Global warming is also likely to have a filtering effect on ecosystems, destroying species that are not mobile. There are much data that can be presented to show the likely impact of climate change on species of living organisms but an interesting one concerns climate range. Each species has a unique range, the set of locations which is where members of that species are found on Earth. A species' range depends on the biotic (living) and abiotic (nonliving) conditions it needs for survival and on geography. Loss of climatic range, which is happening, will lead to loss of species, and there can be significant losses up to a two degree rise in temperature, but which increases dramatically above two degrees.

Water Shortages

The world is already short of water. Two billion people do not currently have safe drinking water and 3.6 billion lack access to safely managed sanitation (BMJ 2023). About 10 percent of the global population already lives in countries with high or critical water stress.

As we have already seen, the global population is set to increase for many decades to come and this will create yet more demand for water. However, climate change can seriously compromise the availability of water. Climate change influences when, where, and how much precipitation falls. It also leads to more severe weather events over time. Also, increasing global temperatures cause water to evaporate in larger amounts, which will lead to higher levels of atmospheric water vapor and more frequent, heavy, and intense rains in the coming years.

There are several implications of this.

- It will lead to more floods since more water will fall than vegetation and soil can absorb.
- The excess water will drain into nearby waterways, picking up contaminants like fertilizer on the way.
- Excess runoff eventually travels to larger bodies of water like lakes, estuaries, and the ocean, polluting the water supply and limiting water access for humans and ecosystems.
- Freshwater glaciers around Earth begin to melt at an unsustainable rate, which results in rising sea levels. With rising of sea levels, salt water can more easily contaminate underground freshwater.

- In the Northern Hemisphere, where snow, a freshwater source typically accumulates, warmer temperatures mean less snowfall, which leaves less water available in local reservoirs after winter. This negatively impacts farmers, who are left without enough water to irrigate their crops in the growing season.

By 2050, the number of people in cities facing water scarcity is projected to nearly double from 930 million people in 2016 to up to 2.4 billion. Urban water demand is expected to increase by 80 percent by 2050. In some parts of the world, especially the Middle East, arguments over access to existing water sources such as rivers could lead to armed conflict.

Food Shortages

The world's food system is, today, very complex. In the distant past, countries would have generated most of the food needed by their populations within their borders. Subsequently, food production has changed over time from a local farm to a global corporate one replete with imports, exports, and shipments crossing oceans and continents every day. This meant that consumers had a much greater choice over foodstuffs than they had historically, and food might also be cheaper when purchased from another country rather than produced internally.

Today, the world's food system can be characterized by two key factors.

- A rich world which vastly consumes more food than it really needs leading to high levels of obesity and related health problems. For example, in the USA, 37.3 percent of adults were found to be obese compared to 6.6 percent in China. Also, large amounts of food amounts are wasted in rich countries.
- Over 800 million people in the world are affected by hunger with around 2.3 billion people in the world (29.3%) being moderately or severely food insecure mainly in poor countries.

Now, the reality is that there is probably enough food in the world to keep everyone fed and to satisfy their needs. However, food is not shared out between countries according to population needs. Food distribution is an outcome of the market for food involving buyers, sellers, and prices. The rich countries have the money to buy more food and at higher prices compared to poorer countries.

Now war, economic shocks, soaring fertilizer prices, and climate change are combining to create a food crisis of unprecedented proportions. Looking ahead a number of problems can be seen:

- The global population is growing.
- Most countries are seeing food inflation of more than 5 percent, and governments are responding by trying to insulate themselves from these shocks.
- Rising export curbs are happening. Countries are starting to ban or restrict food exports because of their own domestic needs for food. The Russia–Ukraine war has severely disrupted food supply chains. Ukraine is a major producer of some of the world's most vital crops, and the conflict has virtually cut off its links to the global food supply.

There are now some alarming trends for global food security. Crop yields have declined and will probably continue to decline as the planet warms and water becomes scarce (Lobel and Field 2007). It is suggested that a third of global food production will be at risk by the end of the century if GHG emissions continue to rise at their current rate.

As climate change worsens, protectionism will probably grow and as well as more food export bans, we may see a reconfiguration of food trading patterns between countries as they prioritize national interest and focus on food security. Climate change increases the risks to food production and distribution, so countries will move toward more trade agreements with proximate countries (even if prices are higher) and/or countries with a shared ideological standpoint. The UAE has said that India will export wheat again but for the Emiratis' consumption only. The two already have an existing trade pact that cuts tariffs on all goods to increase their annual trade to $100 billion within 5 years. Russia and China may become ideological partners in a wide range of areas including oil and food.

Overall, it seems reasonable to assume that climate change will have the following impacts.

- Higher food prices.
- Less choice of food product due to more localized buying.
- Insufficient Food Supply in the World Leading to Widespread Famine in Poorer Parts of the World (See WFP 2022).

Mass Migration

In recent years, we have seen waves of migration into Europe (from Africa and Asia) and to the USA (from South and Central America) some of which is legal and some which is illegal. Some of this migration has been a consequence of war, oppression, and violence in migrant home countries but some of it is just economic migrants, being people seeking a better life in richer countries. However, even migration on this limited scale causes a host of social, economic, and political problems in the recipient countries.

As a consequence of climate change, developed countries can expect future waves of migration well in excess of what has been seen recently. Millions will be fleeing desertification of land, flooding, food shortages, extremely high temperatures, and wars over conflict about things like water. One estimate from the World Bank (2021) estimated that by 2050, some 216 million people could be forced to move from their current homes to somewhere else.

Other models (Wired 2023) have suggested that for every degree of temperature rise, a billion people will be displaced. Initially, most of these displaced people will move to a different part of their home country but this influx of people into different parts of the same country may worsen living conditions in what are often already poor countries. Also, with some smaller countries, it may be that there is little land and resources in the country which is unaffected by climate change. Hence, it is extremely likely that many internally displaced people will try to get to travel to Europe or North America.

Migration rates in recent years will seem tiny compared to the numbers of likely climate change migrants seen in the future. It should be clear that migration numbers of this magnitude just cannot be dealt with by allowing every migrant who has fled their home country to enter Europe or the USA—this is an impossible task. Instead, there will be a need for expanded cities, and entirely new cities that could be built on the habitable fringes of Europe, Asia, and North America, and this will require huge financial resources. At the time of writing, there seems to be no consideration of these sorts of issues.

Armed Conflict

It is probably the case that climate change does not directly cause armed conflict, but that it may indirectly increase the risk of conflict by exacerbating existing social, economic, and environmental factors. Such conflicts could occur between two or more countries or could be internal and thus a civil war.

There are a number of issues at play here.

- **Food:** The production of food may be impacted by climate change such that the yields are reduced, and less food is available to eat.
- **Water:** I have already touched on the fact that there is already a water shortage in many parts of the world, and this will worsen with global warming.
- **Migration:** I have already discussed the situation that climate change may result in hundreds of millions of people leaving the homelands because it has become uninhabitable, and they have moved to another part of their country or to another country. This could easily lead to armed conflict.
- **Grazing:** When cattle herders and agricultural farmers are pushed to share diminishing farm and grazing land due to a changing climate, this can cause conflict. Historians will argue that the root cause of 1984 Rwandan genocide, which killed 800,000 people, was the increased demand and reduced supply of quality arable land in the 1980s.
- **Instability:** The reality and the tragedy are that many of the countries which are most vulnerable to global warming are

already politically and socially unstable. Examples here are Afghanistan, Sudan, Haiti, and so on.

Inter-Generational Equity

Inter-generational equity covers the need to be fair or equitable between various groups in our society. In recent years, it has been noted that in many or most developed countries, there is considerable unfairness to younger people. In the UK, many of the so-called baby boomers had the advantage of free university places, free health care, and a final salary pension scheme. As they get older, they will be placing increasing burdens on the health and social care system and extracting pensions well above what they have contributed to the pension schemes. The only people who can pay for this are our children and grandchildren, who face university loans, high mortgage payments, and inadequate pensions as well as the cost of financing and repaying ballooning government debt. Every year since 2008, governments around the world have borrowed extensive sums to finance the welfare state, of which older generations consume a disproportionate share, while leaving the burden of that debt to future generations. Opinion polls show that children (and their parents) expect future generations to be poorer than their own parents and looking at economic forecasts, this seems a reasonable opinion (Klumpes and Prowle 2013).

On top of this, we now have climate change. As a consequence of overconsumption and abuse of the environment by previous and current generations, it will fall to future generations to pick up the pieces. This could mean all of the problems discussed above plus probably higher levels of taxation to pay for it. This could have been so different if the baby boomer generation (including me) had behaved more responsibly.

Possible Responses

If the above impacts do take place, as I believe the evidence suggests they will, the situation could look pretty grim. The responses to this are probably limited and are discussed below.

Major Scientific Breakthroughs

In relating to the task of mitigating climate change, throughout this book, I have always warned about the dangers of relying on a possible scientific breakthrough which either hasn't yet been thought of or has been thought of but is still in the early phases of development.

The first example, which has already been touched on, is controlled nuclear fusion which is different from uncontrolled nuclear fusion being the hydrogen bomb. Clearly, the development of a source of energy from nuclear fusion would make a massive contribution to energy supply as it would provide a carbon-free source of energy from almost limitless fuel but without the problems of disposing nuclear waste material, which is a problem for nuclear fission. Controlled nuclear fusion research has been going on for 60 years with only limited success. However, in 2022, there was a breakthrough. US scientists at the National Ignition Facility at the Lawrence Livermore National Laboratory (LLNL) in California confirmed that they had overcome a major barrier: that of producing more energy from a fusion experiment than was put in. This was the first time this has happened.

This was a very encouraging result, but it is far too optimistic to think that this is the answer to stopping global warming. The Livermore result was the outcome of a research experiment which needs to be refined, repeated, and improved to get a better energy output. The next stage will then be to develop a working commercial fusion reactor, and this could be several decades away, much too late to help get to net zero by 2050. There are also concerns about the availability of fuel. A fusion reactor requires the element deuterium (often referred to as "heavy hydrogen"). This is abundant and can be extracted from seawater. However, fusion reactions also require tritium (often referred to as "super-heavy hydrogen") and that is far less abundant. There are some doubts as to whether sufficient tritium will be available for large-scale fusion energy production (Science).

The second technological breakthrough concerns carbon capture. Some scenarios of the future increasingly assume the use of technologies that can easily and cheaply remove GHGs from the atmosphere. In such scenarios, it assumed that at some point in the future, a widespread effort will be undertaken that utilizes such technologies to remove CO_2 from the atmosphere and lower its atmospheric concentration, thereby starting to reverse CO_2-driven warming on longer timescales. Current approaches to carbon capture are just not economically viable. Today, carbon capture costs in the range of US$600 per ton of CO_2. The world emits approximately 43.1 billion tons of carbon per year that translates to around $25.8 trillion to sequester all of the carbon emitted each year which is unrealistic. Deployment of such technologies at this scale would thus require large decreases in their costs to become viable and this may not be possible. There are other technical problems with this approach including leakage of CO_2 back into the atmosphere. Hence, even if such technological fixes were practical, substantial reductions in CO_2 emissions would still be essential.

The moral of the story about nuclear fusion, and other technologies such as carbon capture, improved batteries, and so on, is not to rely on them to get out of the hole we find ourselves in. These technologies may or may not come to fruition but for the next few decades, but we need to concentrate on the actions needed to mitigate climate changes, which have already been discussed.

Focus on Adaptation Rather Than Mitigation

In the face of climate change, humanity must act on two broad fronts at the same time—mitigation and adaptation.

As time goes on, it may appear increasingly likely that we are going to fail in our goal of net zero by 2050, and temperatures may look likely to rise beyond the totemic figures of 1.5°C or 2.0°C. In this situation, the impacts outlined in the first part of this chapter will also look increasingly likely. In this situation, it may be that governments feel it essential to deal with the problems that are occurring at the time and, in the immediate, future and put more emphasis on adaptation than they have done in the past. This does not mean that mitigation measures will cease, since global warming has to be stopped, but it does mean some shift of emphasis and resources. Whether this is a wise thing to do is debatable, but it does seem likely.

Added to this, something noted earlier, 70 percent of GHG emissions are due to the activities of just 10 countries in the world. If these top 10 polluting countries fail to take sufficient actions, then the actions of the other 190 or so countries will have only limited impact. Consequently, some may question the relative worth of continuing to use scarce resources to mitigate climate change rather than making greater adaptations to cope with the inevitable impacts of climate change.

There are many adaptation responses, but such responses can be categorized fourfold all concerned with reducing risks and exploiting opportunities as follows:

- ***Infrastructural and technological:*** including engineering, built environment, and high-tech solutions.
- ***Institutional:*** covering economic organizations, laws and regulation, government policies and programs, and so on.
- ***Behavioral and cultural:*** covering individuals, households as well as social and community approaches.
- ***Nature-based:*** including ecosystem-based adaptation options.

Some examples of adaptations, in each category, are given in Figure 9.1.

Infrastructural & Technological	• Improved infrastructure to protect against flooding, sea level rise, heatwaves and extreme heat etc. • Managed retreat where people cannot stay in existing locations a retreat to new locations will be needed. • Early warning systems about oncoming incidences of extreme weather
Institutional	• Changes to land planning arrangements • Revised building codes • Climate based insurance policies
Behavioural/cultural	• Changes in diets • Changes in travel patterns and modes of travel • Changes in consumption of goods and services
Nature-based	• Restoration of natural fire regimes to make catastrophic fires less likely, • Protection and restoration of natural and semi-natural areas

Figure 9.1 Possible adaptation measures

Of course, climate adaptation is not an easy thing to do. Such changes will cost enormous amounts of money and will lead to huge disruption. An LSE report (2022) suggested that the total cost of climate change damage to the UK was projected to increase from 1.1 percent of GDP to 3.3 percent by 2050 and 7.4 percent by 2100. The UN Environment Programme (2022) estimated that the global cost of adapting to climate change impacts is expected to grow to $140–300 billion per year by 2030 and $280–500 billion per year by 2050. Thus, it will require huge amounts of investment funding and the technological skills necessary to make the necessary changes.

Although imprecise, these are scary numbers and governments need to start thinking how they will finance the huge amounts involved. Global public debt is already in excess of $70 trillion ($70,000 billion), and many countries are more concerned with reducing that debt than increasing it. Maybe governments need to start thinking about creating some form of national reconstruction funds in order to set aside funds, now, for large-scale reconstruction in the future. While this may be possible in richer countries, poorer countries will probably find the costs of future climate change adaptation to be unaffordable and will require assistance. Unless there are large-scale transfers of funds, technology, and skills to rich countries to poorer countries, then the future for such countries looks grim. Experiences from the COP summits do not provide much optimism about this.

Climate Engineering

Few issues arouse as much controversy in relation to climate change as the application of what might be termed climate engineering or geo-engineering, these being "technical fixes" designed to halt climate change. With climate change accelerating and little being done to curb the GHG emissions, some scientists have resurrected the idea of deliberate large-scale manipulation of the planetary environment.

Attempts at modifying local weather patterns have been tried for many years, often successfully, with the most common approach being cloud seeding to encourage precipitation. However, caution is needed when tinkering with the weather as the following story relates:

The north of China, a naturally dry area, suffered from water shortages due to soaring consumption in recent years. Around 800,000 hectares of farmland were affected by drought. Precipitation was needed.

In Beijing the snowstorm seemed to come out of nowhere. It coated the roofs in a white glaze. The city, caught in the grip of a decade-long drought, had not seen so much precipitation in months. It was anything but normal.

In fact, the storm in February 2009 was the result of a remarkable confluence of cold air, cloudy skies, and 313 sticks of silver iodide fired into the atmosphere by weather engineers hoping to make something out of nothing. The farmers got their precipitation but passengers waiting for around 100 planes in the airport were delayed and, in a few cases, cancelled, traffic was brought traffic to a standstill on a dozen highways and a few people in the mountains were killed I an avalanche. As expected, Chinese officials claimed a success but did the experiment go to plan. Was it expected to be like this. What if the consequence of the snowstorm had been more serious.

This success in tinkering with the weather underscores a growing risk that has not received the serious international debate it deserves. What happens if someone in our ever-warming world decides to tinker with the climate itself?

The list of unintended consequences of human attempts to manipulate nature is long and varied including the following:

- Concrete jungles creating urban heat islands
- Vast oceanic dead zones resulting from fertilizer use on inland agricultural land
- Intentionally introduced species, such as the cane toad in Australia, that then wreak havoc on ecosystems, among others

Whether the idea is to mimic a volcano's cooling impact on climate by continuously pumping sulfates into the stratosphere or to brighten clouds via crewless ships spewing water vapor, possible problems range from shutting down rainfall in certain regions to unilateral declarations of war.

Climate engineering goes further than weather modification in that it takes action on a global basis to impact on the climate and reduce or stop global warming. There are two main approaches. There are approaches which attempt to reduce the amount of climate change produced by an increase in GHG concentrations and there are approaches that try to remove GHGs that have already been released to the atmosphere. These are described below.

- **Sun blocking:** This would seek to block some of the sun's energy not with a giant space mesh, but with tiny particles suspended in the stratosphere or dusted onto clouds to make them reflect more light. It would be quick and relatively inexpensive, but carbon would continue to build up in the atmosphere. If we ever let those sun-stopping particles dissipate, the effect could be the climatic equivalent of opening a shaken-up bottle of carbonated water.
- **Carbon cleaning:** This would scrub the air of carbon pollution. That could involve massive filters and underground pumps, or it could mean seeding the oceans for phytoplankton and planting new forests to inhale carbon. It would be expensive. It would be slow. But it would take direct aim at the problem, slowing or even reversing the build-up of atmospheric carbon that is driving global warming.

There is one more category proposed and that is that we might take heat that exists near the surface of the Earth and stuff it down deep into the ocean. This hasn't been looked at very much. But, in theory, it's another way of altering Earth's surface temperatures. Obviously, there will be risks associated with the application of geo-engineering and these can be classified twofold.

- **Environmental risks:** This concerns the intended or unintended consequences of a climate intervention. We see from the example in China above that these interventions cannot be precisely implemented, and the outcomes may be different to that expected consequences.

- **Political risks:** Local environmental consequences are concerning but not necessarily the biggest concern. Consider the example shown above about the actions of the Chinese government took a different action, which also resulted in unforeseen snow in Beijing. Imagine that the following year heavy and unexpected snow fell in a neighboring country with major consequences for their population. Whether the Chinese intervention was or was not of the cause of the snowfall in the other country, this becomes almost irrelevant if its politicians and population think it is. The other country might claim reparations from the Chinese who deny this is their fault and such disputes can easily get out of hand. Unilateral actions like this might be a bad idea. I think it's likely that as a result of any climate intervention, there will be winners and losers. In a nuclear-armed world, a world with terrorism and where losers have the ability to strike back at winners, the potential for the kind of political or military risk to overwhelm any environmental benefits is very real. There have even been rumors that the military in some countries have already been considering the use of a climate engineering as a weapon of war. The clearest path to environmental risk reduction is GHG emission reductions. If we were already on a trajectory of reducing GHG emissions to achieve net zero, then we probably won't be thinking about geo-engineering. However, in the current situation where it doesn't look like we are on such a trajectory, it is inevitable that we think about what we'll do if things get really bad. Some argue that If Planet Earth is facing a climate "emergency," then we should leave no option for combating it unexplored. How might we do this? Well, we could undertake climate engineering slowly, speeding up if things were looking good and slowing down if bad things started happening. Unfortunately, that is easier said than done. Political consensus is required at all stages, and this will be incredibly difficult to achieve. Just look at the difficulties in the COPs. If we hit a real global crisis, then countries may be frightened into attempting climate engineering as a last resort.

Conclusion

The first part of this chapter makes grim reading. The implications of failing to achieve net zero by 2050 (which I believe to be likely) implies global temperatures rising to 1.5°C and beyond. In turn, this will have a large number of negative effects concerning climate, water, food, biodiversity, and so on.

Scientific and technological breakthroughs are often posited as the means by which we might avoid these negative effects, but the reality is that these breakthroughs might never happen, or, more probably, will happen too late to have an impact on net zero by 2050, or might prove dangerous. I am a great fan of the sci-fi series Star Trek, which has some wonderful story lines. I am amazed how the crew of the SS Enterprise can take actions from their spaceship to reform the geography and climate of planets, which can be made habitable for humans. However, Star Trek is based on three centuries in the future and even then, they don't always get things right.

I suggest that the most likely course of action is a shift from mitigating climate change to adapting its consequences. The Earth may be a less comfortable place to live but more affluent countries will find ways of making adaptations to reduce the discomfort. In poorer countries, I fear this may not be the case and circumstances will be dire and much worse than they already are.

CHAPTER 10

Implications and Actions for Key Sectors of Society

In this book, I have often referred to climate change as the greatest challenge ever to face humanity. The book has aimed to discuss, comprehensively, the various aspects of climate change, such as its nature, the causes, potential solutions, and likely outcomes.

However, up to this point, the book has addressed these various complex issues in a macrosense by considering climate change as a global issue. In reality, climate change is more than a global issue since it is a phenomenon that affects countries individually and many different parts of society in those countries (and the institutions and components involved) in very different ways.

In this important chapter, I look at the implications of climate change for a number of specific sectors which will be impacted by climate change. These are as follows:

- Governments
- Economies
- Employment
- Businesses
- Public services
- Health

The aim of this chapter is to provide a "guide" about dealing with climate change in many different settings.

Implications for Governments

Every country will have some form of government, but the type of government will vary greatly. All will have a head of state (and possibly a head of government), some form of legislature and some system of law courts. As we saw in Chapter 7, some countries will have a form of democratic government while others will be flawed democracies or be some form of totalitarian state. Nevertheless, they will all be affected by climate change.

The Main Roles of Governments

The roles of a government are multifaceted and encompass various responsibilities aimed at ensuring the well-being and prosperity of its citizens. Some key roles of a government include the following:

- ***Protecting national security***: Governments are responsible for safeguarding the nation's security and defense against external threats.
- ***Foreign relations***: A government has the responsibility of liaison and cooperation with the governments of other countries and international organizations such as the UN, World Bank, COPs, and other nations regarding climate change.
- ***Protecting borders***: Governments will have a responsibility to protect the borders of a nation, which would mean ensuring that the movement of goods and people across borders is lawful
- ***Maintaining law and order***: Governments establish and enforce laws to maintain peace, order, and security within society. They also administer justice through the legal system to uphold the rule of law and protect the rights of individuals.
- ***Providing public goods and services***: Governments are responsible for providing essential public goods and services, such as infrastructure, health care, education, public safety, and social welfare programs, that benefit society as a whole.
- ***Regulating the economy***: Governments play a crucial role in regulating economic activities, promoting fair competition, and ensuring consumer protection.
- ***Promoting social welfare***: Governments implement social welfare programs, such as health care, housing assistance,

unemployment benefits, and social security, to support vulnerable populations and reduce inequality within society.

- ***Fostering infrastructure development*:** Governments invest in infrastructure projects, such as transportation networks, utilities, and public facilities, to support economic growth, enhance connectivity, and improve the quality of life for citizens.

Overall, the primary role of a government should involve serving the public interest and protecting the rights of its citizens, by creating an environment that enables individuals, businesses, and other organizations to thrive and contribute to the overall well-being of society. However, the extent to which governments fulfill these roles will vary greatly between countries.

Activities of Governments

So, what do governments actually do to fulfill the roles described above? Well, a long list could be provided but perhaps the key points are as follows:

- **Legislation:** Governments, via the legislative branch, are responsible for developing and passing new legislation in a variety of areas and sometimes repealing existing legislation
- **Developing policies:** Governments are responsible for developing and implementing public policies in a wide variety of areas relating the roles above
- **Policy implementation:** The implementation of agreed policies either directly or via government agencies and other organizations
- **Revenue raising:** Governments need to raise revenue to finance their activities including the delivery of a variety of public services. The main source of such revenues are various forms of taxation
- **Distribution of funding:** Governments have to develop and utilize appropriate methods for distributing the funds raised through tax revenues to the range of government activities

- **Direct provision of services:** In some situations, public services are delivered not by governments themselves but by a variety of government agencies, voluntary organizations or even private companies. However, such services are paid for via government finance.

Impacts of Climate Change for Governments

Overall, climate change poses significant challenges for governments at all levels. While the specific impacts will vary depending on location and context, it is essential that proactive climate change adaptation and mitigation strategies are developed and implemented to ensure the well-being of citizens and the stability of governance systems. The impact of climate change on governments is multifaceted and far-reaching, affecting numerous aspects of their functioning and responsibilities.

Policy Making	Direct Impacts	Environmental
• International cooperation • Domestic policy coordination • Maintaining public trust and legitimacy	• Natural disasters and displacements • Public health threats • Food insecurity • Energy insecurity	• Loss of biodiversity • Sea level rises and flooding • Infrastructure vulnerabilities
Economic	**Social**	**Financial**
• Disruptions to key sectors of economy • Loss of productivity and GDP • Need to promote innovation and technology	• Equity and inequality • Social and political instability • Law and order and policing	• Long term financial planning • Costs of climate change mitigation • Costs of climate change adaptation • Overseas aid • Taxation policies

Figure 10.1 Climate change challenges facing governments

Shown in Figure 10.1 are a number of challenges facing governments as a consequence of climate change, but this is not exhaustive.

Challenge for Governments

All of the above represent significant challenges to virtually every government on Earth although the precise problems will vary from case to case. It is legitimate to ask how governments will actually cope with the huge and complex challenges posed.

As has already been mentioned in Chapter 1, a major issue for governments is that, at the present time, there are also many other various and serious problems which will occupy the time and minds of elected representatives and officials. These are summarized as follows:

- **Superpower competition:** The worsening of tensions between the United States and China on economic and political grounds has implications for many countries.
- **Ideological conflicts:** A clash of political ideologies in many parts of the world particularly in relation to democracy and autocracy.
- **Major wars:** The world is currently beset by two major wars in Ukraine and the Middle East, which have huge impacts across the globe in political and economic terms.
- **The global economy:** Economic growth in many countries and the world as a whole hasn't really recovered since the 2008/09 global recession. Many governments are searching from ways to boost economic growth, but this might prove difficult.
- **Global public finances:** A "time bomb" caused by governments of all persuasions continuing to borrow large sums of money to finance growing public services rather than using tax revenues from current generations.
- **Artificial intelligence:** Increasing concerns around the world about the potential dangers of AI in many areas including that national security.

- **Ageing populations:** To some degree or another, most countries in the world face the phenomenon of ageing populations. This has major social and economic implications particularly for public services.
- **Country instability:** In many parts of the world (particularly the poorer regions), there is considerable instability and insecurity often resulting in civil war. Clearly, this distracts government from dealing with other social and political issues.
- **Pandemics:** The Covid pandemic lasted almost 3 years and caused huge disruption across the globe. Another such pandemic is always possible and must be guarded against.

Dysfunctionality of Governments

There is a strong argument that many countries in the world suffer from having dysfunctional governments who seem unable to meet the challenges they face. At one extreme, the existence of what are termed "failed states" implies countries where governments have lost their ability to fulfill fundamental security and development functions, lacking effective control over their territory and borders.

Some authors, (Stark et al. 2022) writing about democracies, discuss the existence of dysfunctional governments as being those unable to adapt adequately to the demands of a changing environment and thus produce unintended outcomes that harm the democratic political system. Other authors (Prowle, 2022) discuss the high levels of public policy failure in the UK consequent on dysfunctional governments. Clearly, the additional challenges referred to above are going to cause yet more problems for governments trying to deal with the impacts of climate change.

Constraints Concerning Public Policy

In the context of the above, it is illustrative to consider what are the key constraints on governments in developing and implementing public policies regarding climate change and associated matters. The following are suggested:

- ***Capacity:*** This concerns the limitations on the capacity available to policy makers (elected or not) to develop and implement public policies in relation to climate change. Capacity can be seen as an amalgam of several factors including knowledge and understanding, information, time availability and energy.
- ***Culture:*** The attitude of government toward climate change policy will be influenced by a number of factors, which might be seen as cultural. One of these will be the ideological standpoint of a particular government. Other factors that might be seen as nonpolitical include things like the attitude toward entrepreneurialism, devolution of power, innovation and risk.
- ***Resources:*** Governments of different countries will have vastly different levels of resources available to them to address climate change and other issues. Clearly, there will be a big gulf in resource levels between affluent high-income countries and low-income counties, and the latter will be heavily constrained in what they can do. This is the reason why transfers of funds from rich to poor countries will be necessary to avoid disaster, and this is better done early-on rather than waiting until the disaster occurs. While the COPS have discussed this issue, the level of concrete action has been minimal. Another issue, of course, is the extent to which governments are prepared to levy taxes on their citizens and businesses in order to raise revenues that can be utilized by governments.
- ***Freedoms:*** The freedom for governments to initiate public policies and apply resources to them is constrained by a number of factors. Even in the most totalitarian of states, the leader of the country and its governments will still have some sorts of constraints on their actions due to concerns by influential groups in the country. In democracies, the situation is more difficult, and governments are constrained by a number of factors including the views of the electorate, the views of their own political party and the power they have in their legislature to get legislation and funding approved.

In the light of the points made above, what can governments do to facilitate a better discharge of their duties regarding climate change and other matters. This is not easy, but some possible solutions include the following:

- **Devolution:** In some countries (e.g., the UK), the processes of government are highly centralized at the national level while in other countries (e.g., Germany), much of government activity is devolved to the regional government level. This enables national governments to focus on those issues which only national governments can deal with. Thus, one approach to assist national governments in dealing with the challenges of climate change, in a competent way, is to devolve responsibility for other activities to the regional or local level in the country.
- **Bipartisanship:** Democratic government usually implies several political parties competing with one another for power and influence. Bipartisanship can refer to a situation where the major political parties agree about some but not all parts of a political program. Bipartisanship involves trying to find common ground, but there is a debate whether the issues needing common ground are peripheral or central ones. Bipartisanship often occurs in relation to foreign relations but not domestic issues. However, some would argue that the extent of bipartisanship has worsened in recent years (e.g. the United States) and a reversal of that trend would improve government.
- **Re-engineer the government machine:** In the context of the UK, Dominic Cummings, the former Chief of Staff to the Prime Minister, called for changes to how government works, saying there are "profound problems" with how decisions are made. He complained that the civil service employed too many generalists and lacked people with deep expertise in specific fields such as data scientists, software developers, and economists. In their book entitled *Reforming UK Public Policy Through Elected Regional Government*, Prowle (2022) and others argue that the there are many concerns, about the capabilities of the UK

government machinery in relation to policy formulation and implementation. Similar concerns can be found in many other countries.

Implications for Economies

Rightly or wrongly, possibly, the most important aspect of climate change is seen as the implications it has for the economies of individual countries, regions of the world and the world as a whole. As already seen, climate change will have major implications in a number of areas such as forced migration, floods and droughts, damage to buildings, food insecurity, deaths, and so on, but I suggest that it is the impact on their economies that will most tax the leaders of nations.

What Is Meant by an "Economy" and What Are Its Objectives?

Before we start discussing the impact of climate change on the economies of the world, it is, perhaps, important to discuss what is meant by an "economy."

Basically, an economy is the system for deciding how scarce resources are used so that goods and services can be produced and consumed. Resources are things like land, people (who can work or innovate through their ideas), and raw materials. They are seen as scarce because we have unlimited wants but there are not enough resources to produce the goods and services to satisfy these wants.

The economy is important to all of us. What's happening in the economy can affect us and the decisions we make. Our decisions can also influence how the economy is performing. Every time we choose to buy something (or not to), we are affecting the economy.

In simple terms, an economy can be thought of as comprising five main elements:

- The household sector
- The business sector
- The financial sector

- The government sector
- The overseas sector

However, these different sectors within the economy are linked together. This interdependence means that a change in one sector can affect the rest. Historically, economies were seen, largely, as a private matter of little importance to governments. However, in the last century or so, governments have become much more involved in managing the economies of their countries through a variety of means such as monetary policy, fiscal policy, regulation, subsidies, and so on. In most countries today, governments now have a strong involvement in the management of their economies to a lesser or greater degree.

In considering an economy, there must be a view about the objectives of that economy and how its economic performance should be assessed. There are various approaches to this as discussed below.

- ***Size of GDP:*** The most common measure of an economy is probably that of the gross domestic product (GDP) of an economy. The GDP of an economy is the standard measure of the value added and created through the production of goods and services in a whole economy during a certain period. In broad terms, the GDP of an economy might indicate the affluence of a country with high GDPs of an economy implying more affluence and low GDPs, less affluence.
- ***Economic growth:*** Another important issue is whether the GDP of a country is growing or contracting. For centuries, on Earth, the GDP of the planet was largely unchanged. However, over the past few centuries, there have been periods of significant economic growth and some periods of economic downturns. The Industrial Revolution in the 18th and 19th centuries marked a significant turning point in global economic growth. It led to the rapid industrialization of many countries, resulting in increased productivity, urbanization, and overall economic expansion. In modern times, there is almost an expectation that a country's economy will grow each year, and this rate of growth is seen as an indicator of the success or failure of economic policy. The

reasons for this are that economic growth implies more affluence for the population and also that the taxes collected from these more affluent individuals and companies can be used to enhance public services like health, schools, and so on. However, in the last decade or so, in many countries, economic growth has been weak and well below the average growth post World War II. Consequently, the impact of climate change on this already weakened growth is a further concern.

- ***Income/wealth distribution:*** While GDP is a widely used measure of economic performance, it has certain serious limitations that should be considered. GDP does not provide information about the distribution of income and wealth within a country. It is possible for a country to have a high GDP but significant income and wealth inequality, indicating that the benefits of economic growth are not evenly distributed among the population.
- ***Environmental impacts:*** GDP does not consider the environmental costs associated with economic activities. It does not account for the depletion of natural resources, pollution, or the negative impacts on ecosystems. Therefore, GDP growth may not reflect the sustainability or long-term well-being of an economy.
- ***Regional factors:*** One of the issues discussed earlier in Chapter 7 was that of inequalities between different parts of a country or between countries. Thus, it is insufficient for governments to just look at headline figures (such as GDP) for the economy as a whole. Consideration must be given to ensuring some balance of economic performance between different parts of the country to ensure inequalities don't become excessive.

Impacts of Climate Change on Economies

Climate change has, and will continue to have, significant implications for the economies in a number of ways including the following:

- ***Economic costs:*** Climate change can lead to increased economic costs to all parts of society due to extreme weather events, rising sea levels, and changing climate patterns. These events can damage infrastructure, disrupt supply chains, and impact agricultural productivity, resulting in financial losses for businesses and governments.
- ***Employment:*** The economy of a country is a big driver of employment opportunities for individuals. Addressing climate change requires transitioning to a low-carbon economy, which has implications for employment levels in different sectors. This is discussed further below.
- ***Living standards:*** The state of a country's economy will impact on the cost of goods and services available for purchase by the population of that country and the standards of living people enjoy. Changes in policy to mitigate climate change may impact on this.
- ***Government revenues:*** Government revenues, which are raised through taxation on individuals and organizations, are used to pay for a range of public services, such as health, education, defense, and so on. Activities relating to climate change may impact on the level of government revenues collectible and the delivery of public services.
- ***Overseas trade***: To be able to obtain goods and services from another country, countries have to sell their own goods and services to those countries. This is the basis for overseas trade which is such a huge part of the global economy. Trade policies and agreements may also be influenced by climate considerations, such as carbon tariffs or incentives for sustainable practices.

Thus, it is important to consider the economic implications of climate change and to take proactive measures to mitigate risks and seize opportunities in the transition to more sustainable and resilient economies.

Impacts of Climate Change on Economies

The reality is that climate change is inextricably bound to the economy and in talking about this link it is important to understand, at the outset, that there are several aspects to this, and these are considered below. Moreover, this comment is true for individual countries and the world as a whole.

There are two key aspects to consider

- the impact on the GDP of the country
- the cost implications to the country

Economic Impacts of Climate Change Itself

The very existence of climate change itself (e.g., rising temperatures, flooding, and so on) has had and will have direct impacts on the economies of countries around the world. Global warming will continue, irrespective of whether net zero by 2050 is achieved, and this warming will also have an impact on economies.

These impacts are considered twofold:

- ***Impact on GDP:*** Climate change can have significant impacts on GDP, which is a measure of the economic output of a country. These impacts can be both direct and indirect. Direct impacts of climate change on GDP include the destruction of physical assets and infrastructure due to extreme weather events such as hurricanes, floods, and wildfires. These events can lead to property damage, disruption of production and supply chains, and increased costs for disaster response and recovery. The reconstruction and repair activities following such events can temporarily boost GDP, but the overall impact is negative due to the loss of productive assets.

  - Indirect impacts of climate change on GDP are more complex and can be long term. Changes in temperature and precipitation patterns can affect agricultural productivity, leading to reduced crop yields and increased food prices. This can have a negative impact on GDP, particularly in countries where agriculture plays a significant role in the economy. Climate change can also affect other sectors such as tourism, energy, and manufacturing. For example, rising sea levels and increased temperatures can impact coastal tourism and infrastructure, leading to reduced revenues and job losses. Changes in water availability can also affect energy production, particularly hydropower, and increase costs for cooling in manufacturing processes.

- It is important to note that the exact impact of climate change on GDP can vary depending on the region, the level of vulnerability, and the ability to adapt. Some studies suggest that the economic costs of climate change mitigation could be substantial, with estimates ranging from a few percentage points of GDP to more significant impacts on the long term if mitigation and adaptation measures are not implemented.

- ***Cost implications:*** The impacts of climate change on the economy are wide-ranging and can also be both direct and indirect. Direct impacts include increased frequency and intensity of extreme weather events, such as hurricanes, floods, and droughts, which can lead to property damage, infrastructure destruction, and crop failures. These events can result in significant economic losses, including increased costs for disaster response and recovery. Indirect impacts of climate change include changes in agricultural productivity, water availability, and public health. Rising temperatures and changing precipitation patterns can affect crop yields, leading to food price increases and potential food security issues. Changes in water availability can impact industries such as agriculture, energy production, and manufacturing. Additionally, climate change can have adverse effects on public health, leading to increased healthcare costs and productivity losses.

Economic Impacts of Climate Change Mitigation Activities

Measures to mitigate climate change with the hope of achieving net zero have already been discussed including a shift to renewable energies, improved recycling, greater energy efficiency, and so on. The possible economic impacts of these measures are as follows:

- ***Impact on GDP:*** Climate change mitigation efforts can have both positive and negative impacts on GDP. While the immediate costs of implementing mitigation measures may have short-term negative effects on GDP, the long-term benefits might outweigh these costs. Some potential impacts concern the following:
  - *Transition costs*: Implementing climate change mitigation measures, such as transitioning to renewable energy sources or improving energy efficiency, may require significant investments and changes in infrastructure. These upfront costs can temporarily impact GDP growth.
  - *Innovation and technological advancement:* Mitigation efforts can drive innovation and the development of new technologies. This can lead to increased productivity, efficiency, and competitiveness in various sectors, positively impacting GDP in the long run.
  - *Reduced environmental costs*: Climate change mitigation can help avoid or reduce the costs associated with environmental damage and natural disasters. By preventing or minimizing the impacts of climate change, such as extreme weather events or sea-level rise, economies can avoid the economic losses and expenses associated with recovery and rebuilding efforts.
  - *Market opportunities:* The transition to a low-carbon economy can create new market opportunities for businesses. This includes sectors like renewable energy, sustainable transportation, and green infrastructure. These emerging markets can contribute to economic growth and diversification.

- ***Variable Impacts:*** It is important to note that the specific impacts on GDP will vary depending on the nature and scale of the mitigation measures implemented, as well as the overall policy framework and economic context of each country. However, studies suggest that the long-term benefits of climate change mitigation, such as reduced environmental risks and increased innovation, might outweigh the short-term costs, and have a positive impact on GDP.
- ***Cost implications:*** Mitigating climate change involves reducing GHG emissions and implementing strategies to adapt to its impacts. The economics of mitigation involve assessing the costs and benefits of different mitigation measures. Transitioning to renewable energy sources, improving energy efficiency, and implementing carbon pricing mechanisms are some examples of mitigation strategies. The costs of mitigation can vary depending on the specific measures implemented and the timeframe considered. While there may be upfront costs associated with transitioning to cleaner technologies, studies have shown that the long-term benefits of mitigation, such as reduced health costs, improved energy security, and job creation in green industries, can outweigh the costs.

Economic Implications of Climate Change Adaptation

Measures are already taking place to reduce the impact of global warming that has already taken place around the world. Adapting to climate change involves implementing measures to reduce vulnerability and build resilience to the impacts of climate change. These measures can include infrastructure improvements, changes in land use planning, and the development of early warning systems, among others. The economic impacts of adapting to climate change can be both positive and negative. On the positive side, investing in adaptation measures can help reduce the costs associated with climate-related damage and disruptions. For example, building stronger infrastructure that can withstand extreme weather events can minimize property damage and the need for costly repairs. Similarly, implementing measures to protect coastal areas from sea-level rise can help preserve valuable assets and reduce the economic losses associated with coastal erosion and flooding. However, there are huge costs associated with adaptation. Implementing adaptation measures requires financial resources, and the costs can vary depending on the scale and complexity of the measures. For example, upgrading infrastructure to be more resilient to climate impacts can involve significant upfront investments. Additionally, there may be ongoing costs for maintenance and monitoring of adaptation measures. The economic impacts of adaptation can also be influenced by factors such as the timing of implementation, the effectiveness of the measures, and the level of coordination and collaboration among stakeholders. Early and proactive adaptation efforts tend to be more cost-effective than reactive measures taken after significant damages have occurred. It is worth noting that the economic impacts of adapting to climate change are context-specific and can vary across regions and sectors. Assessing these impacts requires careful analysis and consideration of local conditions, vulnerabilities, and available resources.

The economics of climate change, its mitigation and adaptation are complex, multifaceted, and rife with uncertainty. Various economic models and assessments are used to estimate the costs and benefits associated with different scenarios and policy options and a variety of projections are produced. However, we can't get away from the fact that there is considerable uncertainty associated with all of these matters.

Implications for Employment

I have singled this out because it is a matter of great concern in many countries. It is inevitable that all three aspects of climate change discussed above will have significant implications for employment levels in a country. There will inevitably be job losses in many situations.

The most obvious example in developed countries concerns the fossil fuel industries. In such countries, the phasing out of fossil fuels will inevitably lead to a large-scale loss of jobs for those who are employed in this sector whether that be in the extraction of these items or in the electricity generation industry which uses fossil fuels. Another example might be the automotive industry where the large-scale shift to producing electric vehicles might impact on jobs because of vastly different production methods and skill requirements.

On a larger scale, we can think about countries which rely on tourism for employment. Hefty increases the costs of air transport plus extremely high temperatures might mean a huge drop in tourists, which has implications for employment.

However, it may well be the case that climate change can drive the creation of new jobs. Adaptation approaches will also create economic opportunities. Developing and deploying new technologies and practices to adapt to climate change can stimulate innovation and create jobs in sectors such as renewable energy, sustainable agriculture, and water management. These sectors have the potential to drive economic growth and provide long-term benefits.

However, there are two important points to make.

- Firstly, governments need to ensure that the new jobs created are in the same geographic location as where jobs are lost. This is in order to avoid, as happened with coal communities in many countries, where the clousure of coal mines created unemployment black spots since no new jobs were created..
- Secondly, the standard argument against any sort of automation and change in industrial structures is always met, since the time of the Luddites, with the refrain that history shows that new jobs will be created to substitute for old jobs destroyed. While this has always been historically true, we need to be aware that these climate-related changes are taking place at the same time that the development of artificial intelligence, which is also likely to threaten many jobs. In this situation, new jobs may not appear to replace old jobs and so consideration will need to be given to policies such as universal basic income to support those for whom there is no job.

Implications for Businesses

The business sector is a sector already feeling the impacts of climate change and will continue to do so in the future. This sector will vary in size and mix, between countries, but will always be of a significant size especially in developed countries. As well as meeting the needs of people in society, the sector is important in terms of employment and overseas trade.

Climate change has significant implications for businesses across various sectors and a few examples are discussed below under the headings of risks and opportunities.

Potential Risks to Businesses

The risks involved include the following:

- ***Physical damage***: More frequent and severe weather events are likely to damage infrastructure and impact on business operations. These physical risks need to be assessed and managed to ensure continuity and resilience practices.
- ***Supply chain disruptions***: Climate change can disrupt the supply of raw materials, energy, and transportation networks that businesses rely on. This can lead to production delays, higher costs, and shortages of essential goods. In particular, businesses need to assess the vulnerability of their supply chains and develop strategies to mitigate the risks. This may involve diversifying suppliers, investing in climate-resilient infrastructure, and adopting sustainable sourcing.
- ***Labour shortages and productivity*** - for a variety of reasons such as high temeratures, transport disruption etc, climate change may have an impact on labour availabity and productivity
- ***Changing consumer preferences***: As awareness of climate change grows, consumer preferences are changing. As a consequence of concerns about climate change grow, consumers may shift

toward more sustainable products and services. Businesses that don't adapt may lose out to competitors offering greener alternatives.

- ***Reputational damage*:** There is an increasing demand for sustainable and environmental-friendly products and services. Businesses that fail to address these changing market preferences may face reputational risks and loss of market share. On the other hand, businesses that embrace sustainability can gain a competitive advantage and attract environmentally conscious customers.
- ***Financial implications***: Climate change can have significant financial implications for businesses. Increasing insurance costs and stranded assets are some examples. Businesses need to incorporate climate-related financial risks into their risk management and financial planning processes.

Potential Opportunities for Businesses

The opportunities that may present themselves include:

- ***New markets:*** The transition to a low-carbon economy creates enormous opportunities for sustainable products and services such as in renewable energy, energy efficiency, electric vehicles, sustainable building materials, and other climate-friendly technologies.
- ***Innovation and competitive advantage*:** Investing in climate adaptation and mitigation can stimulate product innovation, cost savings, operational efficiency, enhanced resilience, and a competitive edge in a changing market.
- ***Investor support*:** Investors are increasingly looking for companies with strong ESG (Environmental, Social, Governance) credentials, including that addressing of climate change.
- ***Enhanced brand reputation*:** Proactive and transparent action on climate change can improve brand image, attract customers, and boost employee morale.

How Businesses Can Prepare for Climate Change?

There Are a Number of Important Actions to Be Addressed

- **Assess risks and vulnerabilities:** Identify climate-related hazards that could affect your business and supply chains.
- **Mitigation strategies:** Develop mitigation strategies for the company that are strongly focussed on business activities and improvements
- **Innovate and seize opportunities:** Explore new business models, products, and services that support a low-carbon transition.
- **Adaptation strategies***:* develop and implement plans to reduce vulnerability, to invest in resilience and to implement remedial actions, consequent on climate change
- **Collaborate with stakeholders:** Work with suppliers, investors, communities, and policy makers to address the shared challenges of climate change.
- **Finance:** Consider how to finance the costs of implementing mitigation and adaptation strategies including use of existing reserves, insurance, borrowing or the creation of a climate change fund
- **Communicate and report transparently:** Disclose your company's climate-related risks, strategies, and performance to demonstrate accountability.

It must be remembered that climate change is not a distant threat. It's a present reality that businesses need to address to remain competitive and ensure a sustainable future.

Supporting a Regenerative Economy

To navigate these implications, businesses must consider integrating sustainability into their strategies, adopting environmental-friendly practices, conducting climate risk assessments, and engaging in collaborative efforts to address climate change at a broader level. An approach to doing this is outlined below under the heading of "Good Dividends."

In this chapter, we saw earlier that an economy is about the production, distribution, and consumption of goods and services of often scarce resources—these scarce resources can be termed "capitals." Capitals can be understood as stocks of value, and they come in many forms. Additional to finance (the most well understood form of capital), there are at least five others:

1. Human capital: the skills of people.
2. Social capital: collective knowledge and relationships.
3. Reputational capital: trust we hold in people and organizations.
4. Operational (sometimes known as manufacturing or institutional) capital: the know-how to realize value.
5. Natural capital: the resources of the planet which we shall refer to as planet-community capital and reflects the dual relationship of the planet and humanity.

We shall say more on this shortly. But first, we wish to recap the purpose of an economy. As already discussed, presently economies are traditionally and mostly set up to concentrate on growth of GDP at any costs. The damage occurring to our planet suggests that this approach is causing much harm—a form of degenerative economies in which growth diminishes life. In contrast, a regenerative economic orientation has an explicit purpose to enhance planet and people. The arguments of doughnut economics (Raworth 2017), already presented in Chapter 8, reflect such a regenerative notion described as "reducing the impact on the boundary conditions for life, whilst also enhancing the minimum human essentials for life."

The essence of a regenerative business would reflect the idea that if the business flourishes so does humanity. This is because the business enhances, rather than damages the capitals it uses. For example, in the process of the business producing products or services, employed people (human capital) are developed and enriched through their employment rather than exploited and abused which would be degenerative use of capital.

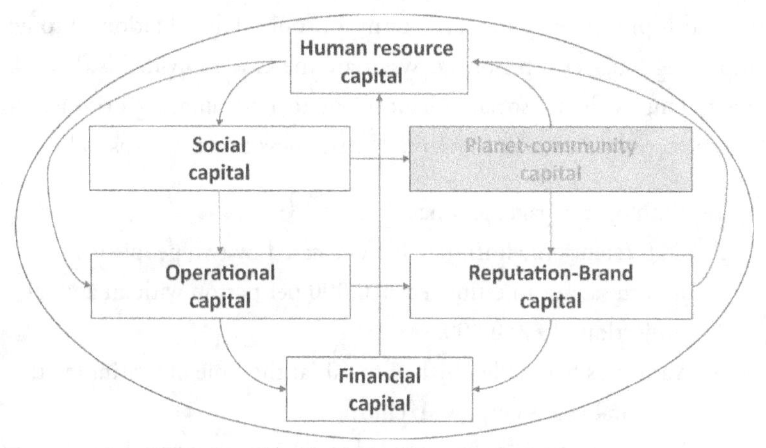

Figure 10.2 A regenerative business

But here's the thing. In a regenerative business, these capitals must operate in a systemic relationship (Kempster et al. 2019). The central thesis is that all six capitals are interconnected and depend on each other to generate what they describe as "Good Dividends"—each capital, including planet-community, generates a dividend for the organization and society (a dividend is the flow of value drawn from the stock of capital). For example, investing in the planet-community creates a dividend to enhance reputational capital but also a dividend with regard to enhancing human resource capital in the form of meaningful work and employee engagement. The dividend from enhanced human resource capital is related to creativity and innovation, productivity, and quality—reflecting operational dividends. With planet-community capital at the heart of the business model, the argument is for a reframing of business and indeed value—the basis for a regenerative business (Kempster and Jackson 2021). Figure 10.2 illustrates the argument:

Using a system of capitals generating good dividends, the idea of growth is reframed to become "good" growth.

This concept of a regenerative business is described below based on a company named Cotteswold Dairy subsequently referred to as CD.

The purpose of CD is to be at the heart of a regenerative and responsible dairy community. They are involved in a partnership with the local prison enabling long-term prisoner rehabilitation through employing long-term prisoners. What are the Good Dividends that CD enjoy along with the social enhancement to rehabilitating prisoners to become active members of society? The business case is thus as follows:

1. Employee engagement has risen.
2. Staff retention: churn rate has decreased by two people per month saving an estimated £10,000 per person with an annual projection of £250,000.
3. Vacancies have fallen by half to 40 (at the time of writing) with measurable less employee "churn."
4. Success of the pilot has now led to six prisoners now being employed across different departments.
5. CD is now working with charities on recruiting homeless people (first jobs back into the workplace) and recovered addicts (chances of employment are as low as convicted sex offenders).

In Chapter 8, I discussed the idea of alternative economic models, and I expressed much caution about the extent to which they could be implemented. However, this above case study shows that if a business can embrace the idea of Good Dividends and become regenerative, then we have the very real opportunity in addressing climate change for business to become part of the solution rather than a major problem.

Implications for Public Services

Earlier in this chapter, it was noted that as well as developing public policies, one of the functions of government is to deliver public services to its people. However, it was also noted that many public services are not delivered directly by government itself but are the responsibility of public agencies, voluntary sector organizations, and even private companies using finance provided by the government. Thus these public services are paid for by public funds transferred to the providers. In this sector, we look at the impact of climate change on these providers.

The implications for public services of climate change are vast and varied. As temperatures rise and extreme weather events become more frequent, public services will face challenges in areas such as emergency response, infrastructure maintenance, public health, and resource management. Governments and public service organizations will need to invest in adaptation and mitigation strategies to address these challenges and ensure the resilience of public services in the face of climate change. Additionally, there may be increased pressure on public services to address issues such as air quality, water scarcity, and natural disaster preparedness.

Overall, climate change poses significant risks to the functioning of public services and will require proactive planning and investment to mitigate its impacts. The following are some key areas.

Disruption and Increased Demand

There are a number of factors here including the following:

Strained Infrastructure

Extreme weather events like floods, heatwaves, and storms can have severe implications in a number of areas including the following:

- ***Roads, bridges, and public transport:*** Damaged infrastructure leading to more frequent repairs, service disruptions, and higher costs for maintenance.
- ***Coastal areas:*** Sea-level rise and storms will damage coastal infrastructure, requiring costly defense measures, relocation of facilities, or even service cessation in vulnerable areas.
- ***Power and water utilities:*** Power outages due to storms, and water shortages due to droughts and contamination will disrupt essential services and strain utility providers.

Social and Economic Disruption

- ***Displaced populations:*** Climate disasters can displace people from their homes, requiring emergency response and resettlement efforts. Populations may need to be relocated due to uninhabitable areas or lost livelihoods, creating pressure on housing, public services, and social systems in receiving areas.

- ***Vulnerable populations*:** The elderly, those with pre-existing conditions, and low-income communities are disproportionately affected by climate change impacts, increasing their reliance on social support services.
- ***Food security*:** Disruptions to agriculture due to weather events and water scarcity could lead to food shortages, increased costs, and social unrest, putting a strain on food banks and welfare services.

Need for Transition and Mitigation

Overall, climate change demands that public services become more resilient, adaptable, and sustainable. This will require investment in infrastructure, new technologies, and training for public sector workers. Some examples are as follows:

- ***Public awareness*:** Public service providers can play a crucial role in educating the public about climate change and promoting environmentally responsible behavior.
- ***Investment in resilience*:** Public service providers must invest in climate-proofing infrastructure, developing robust emergency planning protocols, and upgrading early warning systems. This requires significant financial resources and long-term planning.
- ***Building resilience*:** Expanding mental health services, support programs for vulnerable groups, and strategies to manage displacement will be essential aspects of effective adaptation.
- ***Collaboration and data-driven approach*:** Public services will need to work with climate scientists and other sectors to improve risk assessments, create comprehensive adaptation plans, and efficiently allocate resources.
- ***Energy use*:** Public buildings can become more energy-efficient, reducing GHG emissions, and lowering operational costs.
- ***Sustainable practices*:** Public services can adopt sustainable practices like waste reduction, water conservation, and green transportation to lessen their environmental impact.

AREA	POSSIBLE IMPLICATIONS
Governments	• Need to find addi tional financial resources for health systems • Need to revise health policies
Economies	• Labour shortages due to high levels of sickness • Possible reductions in consumer demand for some products/services • Impacts are worse in poorer countries
Employment	• Increased numbers of persons unable to work • Greater sickness absence
Businesses	• Inability to recruit sufficient skilled staff • Lower levels of productivity
Public services	• Increasing demands for health and social care services • Staff shortages due to staff sickness

Figure 10.3 Possible implications of health status decline

Implications for Health and Health Systems

In this last section, I focus on the implications of climate change for health and health systems. A deterioration in the health status of individuals and society at large will, clearly, impact on the personal lifestyles and contentment of those affected. However, it is not often appreciated that there will also be negative ripples throughout society at large in many ways.

If we just look at the earlier sections discussed in this chapter, it can quickly be seen a deterioration in the health status of a country's population will have implications in other areas as shown in Figure 10.3.

Hence, it is important to consider the potential implications of climate change for the health of a country's population and its health systems. Firstly, let us consider some of the impacts of climate change and the ways they might impact on health status.

- **Heatwaves:** One of the most hazardous climate change risks to health is heatwaves, which are increasingly prevalent. They can cause dehydration, heat exhaustion, and heat stroke. Elderly people, young children, and people with chronic medical conditions are especially vulnerable to heat-related illnesses. Respiratory problems can be exacerbated by air pollution and smog, which are worsened by hotter temperatures and increased ozone levels. Climate change can also lead to the spread of vector-borne diseases, as warmer temperatures, and changing weather patterns create new habitats for disease-carrying insects.

- **Flooding:** Flooding such as flash floods, river floods, or coastal floods can contaminate water supplies with chemical pollutants and increase the risk of waterborne diseases. Flooding also creates conditions that are conducive to the spread of vector-borne diseases. For example, flooding can lead to an increase in the populations of mosquitoes and other insects that carry diseases such as malaria and dengue fever.
- **Cyclones, severe storms, and associated hazards:** Cyclones and other types of storms can cause severe damage to infrastructure and homes, causing significant injuries and fatalities. They can also contaminate water supplies and lead to the spread of waterborne diseases.
- **Drought:** Drought can cause crop failures, which can lead to food shortages and malnutrition. It can also lead to water shortages, which can also increase the risk of waterborne diseases.
- **Mental health:** As well as the impact on physical health, climate change can lead to increased levels of anxiety, depression, and posttraumatic stress disorder. Climate change can also exacerbate existing mental health conditions, and this is a phenomenon that is already occurring.

The impact on health outcomes can also be affected by the readiness and resilience of health systems. However, in turn, the health systems themselves can also be negatively affected by climate change. Healthcare professionals experience the physical and mental health risks of climate change more acutely than the general population. Climate change not only disrupts their lives but also makes their jobs more challenging, raising the risk of burnout. In some countries, storms, floods, wildfires, and other extreme events often prevent them from traveling to healthcare facilities.

Climate change will introduce new risks to health which are not necessarily known to health professionals in a country. Hence, it is vital for the healthcare workforce to be trained in identifying and addressing these risks. In the event of health system disruption, the workforce needs to quickly adapt to delivering care under highly challenging circumstances. When people are displaced, injured, or made ill by climate change, it is mainly up to health systems to deal with the altered and increased health burden. Building resilience, like investing in resources to prepare for climate change and taking steps to reduce healthcare's contribution to climate change, is, therefore, vital to ensuring the U.S. health system can meet the ever-increasing needs of all Americans.

Specifically, there are a number of actions that can be taken in order to reduce the health harm of climate change. These include the following:

- Providing better information and education on the health risks posed by climate change
- Working with other sectors (e.g., businesses) to develop policies and programs that address the health risks of climate change
- Supporting research on the health impacts of climate change
- Monitoring the health impacts of climate change

Conclusions

This chapter has surveyed the implications of climate change for a range of sectors and organizations, which can be found in almost all countries. While climate change is a global issue that requires global responses, this chapters aims to set out the actions that can be taken by individual organization to deal with the challenges it faces from climate change.

CHAPTER 11

Conclusions

Introduction

In the 200 or so pages that comprise this book, I have reflected on and described the issues of climate change and what I think is likely to happen. It is quite clear that the people who initiated the Industrial Revolution would never have dreamed that their activities (including the large-scale use of fossil fuels) would lead to a situation the world finds itself in today and we must wonder where our world is going. In this final chapter, I try to summarize three things:

- Where have we got to?
- What happens next?
- Future actions?

Where Have We Got To?

As discussed in Chapter 4, the headline objective of COP summits and the international community is that many countries have pledged to aim to achieve a target of having net zero GHG emissions by 2050. Others have pledged a later date, and some have made no pledges. I have argued in this book that a very wide range of evidence suggests that, for most of the top 10 polluting countries (which account for almost 70 percent of GHG emissions), the achievement of net zero by 2050 is very unlikely to happen for a wide range of reasons. At some time in the next decade or so, it will probably be accepted that net zero by 2050 is unattainable and new plans will need to be made.

In light of this, we must consider the current situation under the headings below.

Greenhouses Gas Trends

Emissions of GHG have risen rapidly in the second half of the 20th century and the 21st century. Leaving aside the blip caused by the Covid pandemic, levels of GHG emissions are still rising. In May 2023, new data from Hawaii's Mauna Loa Observatory showed that average monthly carbon dioxide (CO_2) levels reached over 420 parts per million the highest on record and 50 percent higher than in pre-industrial levels (UC San Diego 2023).

Before levels of emission can fall and make progress toward net zero, they first have to stop rising. There is some evidence that the rate of increase may be slowing and GHG emissions could plateau in the next few years. However, there is very strong and increasing doubt as to whether the emissions will plateau and fall quickly enough to avoid a 1.5°C or even 2.0°C temperature increase.

Temperature Trends

The surface temperature of the Earth has been warming over a period of years and this seems likely to continue into the future. There is now a growing consensus that we will fail to keep the rise in the Earth's temperature to 1.5°C and some are saying this could happen before the end of the decade. This will have serious implications for life on Earth. We may even struggle to keep the rise within 2.0°C. However, even if emissions stopped rising, it would take many thousands of years for atmospheric CO_2 to return to "pre-industrial" levels due to its very slow transfer to the deep oceans and ultimate burial in ocean sediments. Hence, surface temperatures would remain elevated for at least a thousand years, implying a long-term commitment to a warmer planet due to past and current emissions.

Weather and Climate Trends

In recent years, we have noted, with our own senses, the growth in extreme weather situations such as record high temperatures, record low temperatures, extended droughts, record floods, and stronger storms.

This can't be a coincidence and these trends must be linked to climate change and such events are set to continue or worsen as the Earth warms up more. However, the incidence of such extreme weather events is unlikely to be spread evenly across the globe and some areas will be affected much worse than others.

Biodiversity/Habitat Trends

The loss of biodiversity on Earth and its complex relationship with climate change and habitat destruction was discussed at some length in Chapter 5. It will suffice to say that consequent on climate change, we can expect further bad news in these areas with incalculable implications for life on Earth.

Hardship Trends

The term hardship is a rather vague and relative term. To someone living in high-income countries, hardship might mean things like having to queue in airport departure lounges, having to be pay more for energy or having to put up with a smaller range of foods. In poor countries, hardship might mean walking five miles for fresh water, wondering where food is going to come from or having to flee your home because of desertification of the land.

I am convinced that while people in rich countries may suffer hardships because of climate change, and the attempts to mitigate it, this hardship pale into insignificance compared to the impact on poor countries. While rich countries will probably be able to spend billions or trillions of dollars to adapt to climate change, poor countries will just not have the resources available to do this. I fear we will be seeing large-scale famines, droughts, disease epidemics (which may spread to rich countries), and mass migration beyond anything we have seen in modern times.

Internal Politics and International Relations Trends

I suspect the ongoing issues of climate change will impact, negatively, on the internal politics of a nation. Virtually every week, one can read something in the media which implies backsliding by one country or another concerning its pledges in relation to climate change. Just a few examples are as follows:

- The European Union has a policy applicable to all members that it will stop manufacturing petrol/diesel driver cars as from 2035. It has recently emerged (The Economist 2023 (1)) that the German government is trying to negotiate an opt out to delay the implementation of this policy in Germany following lobbying from the German automotive industry which is the largest in Europe.
- In 2021, the COP summit claimed that they were consigning coal to the ash heap of history. Governments promised to stop building coal-fired power plants, and financiers pledged to stop financing coal mines. It isn't happening (The Economist 2023 (2)). Russia's invasion of Ukraine pushed coal consumption to record levels in 2022 but although this shock has faded, global coal demand is still set to rise a little this year. If global temperature is to be limited to 1.5°C, coal production was supposed to fall by two-thirds by the end of the decade but is now projected to fall by less than a fifth.
- Despite the importance of reducing deforestation, given the role of trees as a carbon sink, about 16,000 sq. miles of tropical rainforest was lost in 2022, most of it destroyed to make way for cattle and commodity crops. Brazil accounted for 43 percent of the loss, with Democratic Republic of the Congo and Bolivia responsible for about 13 and 9 percent, respectively.

I think we can expect these sorts of regressive actions to continue.

There will also be impacts on relationships between countries. Particularly in poorer parts of the world with tensions arising from matters such as food shortages, water access and migration which may well lead to conflict.

It is clear that effective mitigation of climate change requires strong international cooperation and collaboration. While a huge amount of research has been undertaken, the activities and outcomes of the various COP summits do not appear to have produced the changes needed to halt global warming and climate change in relation to planned objectives. Furthermore, the current state of international relations in the world seems unlikely to improve this situation.

What Happens Next

Several decades ago, there was an alternative approach available to Governments and that was to grasp the nettle earlier and stronger and to take radical actions, to counter global warming. As it is, we have left it too late. As the situation worsens, and technological breakthroughs are not achieved in time, the situation will become more and more difficult.

Ssubject to ongoing research and analysis, I suggest that we can be reasonably certain about four things:

- That the future costs and benefits associated with climate change are very uncertain but will be unevenly distributed across the globe.
- In broad terms, the costs of dealing with the impacts of climate change will fall disproportionately on developing countries, while the financial costs of cutting GHG emissions, to mitigate those impacts, will fall mostly to developed nations.
- Even in developed countries, it will probably be the case that the most affluent will have the resources to deal with the effects of climate change (e.g., air conditioning, cold pools, electric cars, and so on) but the less affluent will probably be in for an uncomfortable time.
- While the impacts of climate change on developed countries will be uncomfortable and disruptive, the impacts on poor countries will be catastrophic with immense suffering. As Table 7.1 showed, such countries are most likely to be more vulnerable to climate change, less well prepared, and lacking in the

resources to counter it. Just one example concerns the numbers of climate change migrants in the world. Future numbers will dwarf, by perhaps a hundredfold, the numbers currently trying to enter Europe and North America. It is clearly impossible to bring them all to sanctuary in Europe or the North America but as noted earlier in the book, what needs to happen is the construction of new towns and cities on the fringes of Europe, North Africa, North America, and so on. This will require rich countries to put forward mega-sums of money for this purpose.

Future Actions

In the light of the above, it is important to consider what actions might be taken, in future, by various actors.

International Level

Presumably COP summits will continue to be the main international forum for the foreseeable future and much interesting and important work will be undertaken prior to the annual events. However, whether the effectiveness of the COPs in achieving stronger actions to counter climate change remains doubtful. The top 10 polluting countries would need to commit to much stronger actions than they have in the past and given the nature of many of those 10 countries this does not seem likely.

National Level

At a national level, governments of developed countries will probably continue to pretend that they are aiming for the objective of net zero by 2050 (or a later date) although the chances of this happening are slim and there must come a point where they will admit this. In the meantime, I suspect they will be keeping one eye on the international (COP) situation while keeping the other eye on the impacts of their mitigation actions on their own economies. This may impact on the emphasis they give to mitigation and the mount they invest. I would also hope that they will start to give greater attention to the tasks of adapting to the impacts of climate change including the creation of a fund to finance the measures that will be required in future years. They should also start to prepare their populations for some of the uncomfortable and disruptive changes that might need to be made in their society's consequent on climate change. This could include such things as travel patterns, working patterns, purchasing patterns, and so on.

In the developing countries, the situation is very different. Many of them face impacts of climate change which are beyond their capabilities (and resources) to adapt and deal with. They probably need to work together to lobby the richer countries for assistance in this area, pointing out the implications if this does not happen.

Sectors

Chapter 10 deals with, at some length, the sorts of actions that different sectors (e.g., governments and businesses) in society might undertake to counter the impact of climate change on them. Such actions should be applied and continued irrespective of what happens internationally or within their own country. However, they will, of course, need to refine their actions in the light of what is happening nationally and internationally.

Individuals and Communities

Throughout the world and in many countries, there are millions of individuals and communities taking a variety of actions to reduce the personal impact they are having on the climate. Such actions are to be applauded and encouraged among others. If sufficient numbers of individuals and communities took such actions, then it is likely that, in turn, this would affect the standpoint of their governments.

Menion should b made here of the ripple effect. This demonstrates that the potential power of individual actions to create meaningful and lasting change. Whether it's a small act of kindness, a commitment to sustainability, or a stand for justice, every action has the potential to create a ripple that extends far beyond its initial impact.

References and Sources

"A Fragile Recovery." 2023. OECD. www.oecd.org/economic-outlook/march-2023/.

"Accelerated Dryland Expansion Under Climate Change." 2016. Climate Change Post. www.climatechangepost.com/news/2016/3/5/accelerated-dryland-expansion-under-climate-change/

"Another Year of Extreme Jeopardy for Those Struggling to Feed Their Families." 2022. World Food Programme(WFP).

"Around the World, These Are Just Some of the Cities at Risk of Rising Sea Levels." 2021a. WEF. www.weforum.org/agenda/2021/12/coastal-cities-underwater-climate-change/.

"Broken Record: Atmospheric Carbon Dioxide Levels Jump Again. 2023. UC San Diego. https://today.ucsd.edu/story/broken-record-atmospheric-carbon-dioxide-levels-jump-again.

"Carbon Footprint by Country." 2022. World Population Review. https://worldpopulationreview.com/country-rankings/carbon-footprint-by-country.

"Climate at a Glance Global Time Series." 2023. NOAA. www.ncei.noaa.gov/access/monitoring/climate-at-a-glance/global/time-series/globe/land/1/6/2000-2023.

"Climate Change Could Force 216 Million People to Migrate Within Their Own Countries by 2050." 2021. World Bank. www.worldbank.org/en/news/press-release/2021/09/13/climate-change-could-force-216-million-people-to-migrate-within-their-own-countries-by-2050.

"Climate Change: The Physical Science Basis." 2021. IPCC. www.ipcc.ch/report/ar6/wg1/downloads/report/IPCC_AR6_WGI_SPM_final.pdf.

"Climate Target Update Tracker." n.d. Climate Action Tracker. https://climateactiontracker.org/.

"CO_2 and GHG Emissions Country Profiles." 2023. Our World in Data .

"Conference of Parties List." n.d. Climate Change. www.downtoearth.org.in/climate-change/coplist.

"Countries by GDP." 2024. PopulationU. www.populationu.com/gen/countries-by-gdp.

"Country Populations." n.d. Population Pyramid www.populationpyramid.net/population-size-per-country/2022/.

"Economy for the Common Good." n.d. ECG. www.ecogood.org/.

"Encyclopaedia of Philosophy—The Prisoners Dilemma. 1997. Stanford University. https://plato.stanford.edu/entries/prisoner-dilemma/.

"Feeling the Heat." 2021a. WWF. https://www.wwf.org.uk/sites/default/files/2021-06/FEELING_THE_HEAT_REPORT.pdf

"Forests and Deforestation." n.d. Our World in Data. https://ourworldindata.org/forests-and-deforestation.

"Germany Is Letting a Domestic Squabble Pollute Europe's Green Ambitions." 2021a. The Economist. www.economist.com/europe/2023/03/09/germany-is-letting-a-domestic-squabble-pollute-europes-green-ambitions.

"Gini Index." n.d. World Bank. https://data.worldbank.org/indicator/SI.POV.GINI.

"Global Climate Change: Vital Signs of the Planet." 2024. NASA. https://climate.nasa.gov/vital-signs/sea-level/?intent=121.

"Global Peace Index." 2022. www.visionofhumanity.org/wp-content/uploads/2022/06/GPI-2022-web.pdf.

"Global Warming Petition Project." n.d. www.petitionproject.org/.

"History Base of the Global Environment." n.d. www.pbl.nl/en/hyde-history-database-of-the-global-environment.

"How 'Shared Socioeconomic Pathways' Explore Future Climate Change." 2018. Carbon Brief. www.carbonbrief.org/explainer-how-shared-socioeconomic-pathways-explore-future-climate-change/.

"How to Stop Overpopulation? 5 Possible solutions to this growing issue. 2021. Tomorrow.City. https://tomorrow.city/a/overpopulation-solutions.

"IEA Sees Global Energy Emissions Peaking in 2025." October 27, 2022. The Economic Times.

"If Emissions of GHGs Were Stopped, Would the Climate Return to the Conditions of 200 Years Ago?" 2020. The Royal Society. https://royalsociety.org/topics-policy/projects/climate-change-evidence-causes/question-20/.

"Impact of 4°C Global Warming." n.d. Green Facts. www.greenfacts.org/en/impacts-global-warming/l-2/index.htm.

"Impact of Climate Change at 1.5°C, 2.0°C and Beyond." n.d. Carbon Brief. https://interactive.carbonbrief.org/impacts-climate-change-one-point-five-degrees-two-degrees/

"Impacts of 1.5°C and 2.0°C Warming." 2018. Climate Council Office. www.climatecouncil.org.au/wp-content/uploads/2021/04/Infographic-page-35-scaled.jpg.

"India's Ambitious 2070 Zero Emission Target Needs $10 Trillion Investment." 2022. India Times. https://energy.economictimes.indiatimes.com/news/renewable/indias-ambitious-2070-zero-emission-target-needs-10-trillion-investment/96512902.

"IPCC Report: Sea Level Rise Is a Present and Future Danger." 2019. NRDC. www.nrdc.org/bio/rob-moore/ipcc-report-sea-level-rise-present-and-future-danger.

"Is India's Pledge of Net Zero by 2070 an Ambitious Target—or Worthless Words?" November 5, 2021. The Guardian.

"Land Area-Countries Compared." n.d. Nationmaster. www.nationmaster.com/country-info/stats/Geography/Land-area/Square-miles#google_vignette

"Mapped: How Climate Change Affects Extreme Weather Around the World." 2022. Carbon Brief. www.carbonbrief.org/mapped-how-climate-change-affects-extreme-weather-around-the-world/.

"Mass Climate Migration Is Coming." 2023. Wired. www.wired.co.uk/article/climate-migration-environment.

"Missing Logic of Australian Prime Minister's Denial of Climate Change Link to Bushfire. October 29, 2013. The Guardian.

"ND Gain- Country Index." 2022. University of Notre Dame. https://gain.nd.edu/our-work/country-index/rankings/.

"Net Zero Emissions Race." 2022. ECIU. https://eciu.net/netzerotracker.

"Net-zero Germany: Chances and Challenges on the Path to Climate Neutrality by 2045." 2021. McKinsey's. www.mckinsey.com/capabilities/sustainability/our-insights/net-zero-germany-chances-and-challenges-on-the-path-to-climate-neutrality-by-2045.

"Number of Cars in the UK 2023." 2023. NimbleFins. www.nimblefins.co.uk/cheap-car-insurance/number-cars-great-britain.

"Number of Planes in Air Will More Than Double by 2038. 18 September 2019. www.independent.co.uk/travel/news-and-advice/planes-flying-forecast-aviation-airbus-market-a9110136.html.

"One Planet, How Many People? A Review of Earth's Carrying Capacity." 2012. United Nations Environment Programme(UNEP). https://na.unep.net/geas/archive/pdfs/GEAS_Jun_12_Carrying_Capacity.pdf.

"Out of Gas." 2022. Science. www.science.org/content/article/fusion-power-may-run-fuel-even-gets-started.

"Overseas Travel and Tourism: 2020. 2021. ONS. www.ons.gov.uk/peoplepopulationandcommunity/leisureandtourism/articles/overseastravelandtourism2020/2021-05-24.

"Polar Bear Population." 2021b. World Wildlife Fund. www.arcticwwf.org/wildlife/polar-bear/polar-bear-population/.

"Population by Age Group." n.d. Our World in Data. https://ourworldindata.org/grapher/population-by-age-group-with-projections?country=-CHN.

"Population Projections by countries." n.d. INED. www.ined.fr/en/everything_about_population/data/world-projections/projections-by-countries/.

"Poverty Update." 2022. World Bank. https://blogs.worldbank.org/opendata/april-2022-global-poverty-update-world-bank.

"Poverty Update." 2022. World Bank. https://blogs.worldbank.org/opendata/april-2022-global-poverty-update-world-bank.

"Road Traffic By VehicleType." n.d. Department for Transport. https://roadtraffic.dft.gov.uk/summary.

"Taking Action for the Health of People and the Planet. n.d. www.un.org/en/climatechange/science/climate-issues/health.

"The Commons Movement." n.d. Schumaker Centre. https://centerforneweconomics.org/apply/the-commons-program/movement/.

"The Impact of Climate Change on Our Health and Health Systems." 2022. The Commonwealth Fund. www.commonwealthfund.org/publications/explainer/2022/may/impact-climate-change-our-health-and-health-systems.

"The Sky Is Falling. 2024. Real Climate. www.realclimate.org/index.php/archives/2006/11/the-sky-is-falling/.

"The Struggle to Kill King Coal. 2023b. The Economist. www.economist.com/leaders/2023/06/08/the-struggle-to-kill-king-coal.

"This Is How Climate Change Could Impact the Global Economy." 2021. World Economic Forum. www.weforum.org/agenda/2021/06/impact-climate-change-global-gdp/.

"This Is What 3°C of Global Warming Looks Like." 2021. The Economist. www.economist.com/films/2021/10/30/this-is-what-3degc-of-global-warming-looks-like.

"What Are Mass Extinctions, and What Causes Them?" 2019. National Geographic. www.nationalgeographic.com/science/article/mass-extinction.

"What Did the UK's Electricity Generation Mix Look Like in 2022?" n.d. EDIE. www.edie.net/what-did-the-uks-electricity-generation-mix-look-like-in-2022/.

"What Is Post-Growth Economy and Why Is It Necessary. n.d. Post Growth Institute. www.postgrowth.org.

"What Will Climate Change Cost the UK? Risks, Impacts and Mitigation for the Net-Zero Transition. LSE. www.lse.ac.uk/granthaminstitute/publication/what-will-climate-change-cost-the-uk/.

"When Might the World Exceed 1.5°C and 2.0°C of Global Warming?" 2020. Carbon Brief. www.carbonbrief.org/analysis-when-might-the-world-exceed-1-5c-and-2c-of-global-warming/.

"Which Form of Transport Has the Smallest Carbon FootPrint?" 2023. Our World in Data. https://ourworldindata.org/travel-carbon-footprint.

"World Electricity Generation." 2021. World Energy Data www.worldenergydata.org/world-electricity-generation/.

Adapatation Gap Report. 2022. UN Environment Programme. www.unep.org/resources/adaptation-gap-report-2022.

Anand, S. 2018. "The Curse of Optimism." www.visionofhumanity.org/wp-content/uploads/2022/06/GPI-2022-web.pdf.

Attenborough, D. 2018. In *Life on Earth: The Greatest Story Ever Told.* Collins.

BMJ. March 22, 2023. "Governments Must Speed up Action to Enable Access to Clean Water for All, Say UN Bodies." *British Medical Journal.*

Chalmers, D. 2016. "The Hard Problem of Consciousness." www.organism.earth/library/document/hard-problem-of-consciousness.

CIA. n.d. "GDP—Composition, By Sector of Origin." The World Factbook. www.cia.gov/the-world-factbook/field/gdp-composition-by-sector-of-origin/.

Claessens, M. and M.A. Kose. n.d. "Recession: When Bad Times Prevail." International Monetary Fund. www.imf.org/external/pubs/ft/fandd/basics/recess.htm.

Climate Zones of the World Map. n.d. www.worldmap1.com/climate-zones-of-world-map.

Cook, J. July 22, 2015. "The 5 Tell-Tale Techniques of Climate Change Denial." CNN. https://edition.cnn.com/2015/07/22/opinions/cook-techniques-climate-change-denial/index.html.

Cook, J., D. Nuccitelli, S.A. Green, M. Richardson, B. Winkler, R. Painting, R. Way, et al. 2013. "Quantifying the Consensus on Anthropogenic Global Warming in the Scientific Literature." Environmental Research Letters. https://iopscience.iop.org/article/10.1088/1748-9326/8/2/024024/pdf.

Crunden, E.A. 2021. "How Useful Is Recycling? Really?" The Atlantic. www.theatlantic.com/science/archive/2021/01/recycling-wont-solve-climate-change/617851/.

Desjardin, J. 2019. "$69 Trillion of World Debt in One Infographic." www.visualcapitalist.com/69-trillion-of-world-debt-in-one-infographic/.

Fukuyama, F. 1992. "The End of History and the Last Man." Free Press. ISBN-978-0-02-910975-5.

Global Average Temperature Records. n.d. The Met Office www.metoffice.gov.uk/weather/climate/science/global-temperature-records.

Green, D. 2021. "World Inequality Report." https://frompoverty.oxfam.org.uk/world-inequality-report-2022-a-treasure-trove-of-trends-and-new-data/.

Howarth, R.B and K. Kennedy. January 2016. "Economic Growth, Inequality, and Well-Being." *Ecological Economics* 121, pp. 231–236. www.sciencedirect.com/science/article/abs/pii/S0921800915004024#:~:text=Although%20income%20growth%20is%20often%20interpreted%20as%20a,well-being%20and%20material%20prosperity%20%28Daly%2C%20-1977%2C%20Victor%2C%202008%29.

International Churchill Society. 2016."Special Report on Climate Change and Land." 2019. IPCC. www.ipcc.ch/site/assets/uploads/2019/11/SRCCL-Full-Report-Compiled-191128.pdf.

Kamark, E. 2019. "The Challenging Politics of Climate Change." www.brookings.edu/articles/the-challenging-politics-of-climate-change/.

Kempster, S, T. Maak, and K. Parry. 2019, In *The Good Dividends: A Systemic Framework of Value Creation*. Routledge.

Kempster, S. and B. Jackson. January 2021. "Leadership for What, Why, for Whom and Where? A Responsibility Perspective." *Journal of Change Management* 21, no. 3, pp. 1–21.

Kim. 2022. "Assessing South Korea's Transition to Net Zero." https://carnegieendowment.org/2022/11/22/assessing-south-korea-s-transition-to-net-zero-pub-88426.

Klumpes, P. and M.J. Prowle. December 2013. "What Lies Beneath." *Public Finance.*

Lobel, D. and C.B. Field. 2007 "Global Scale Climate–Crop Yield Relationships and the Impacts of Recent Warming." *Environmental Research* 2, no. 1.

MacKinnon, J.B. 2021. In *The Day the World Stops Shopping*. Bodley Head.

Mallender, J. 2022. "The Health Economics of Climate Change." www.economicsbydesign.com/the-health-economics-of-climate-change/.

NASA. 2019. "A Degree of Concern: Why Global Temperatures Matter." https://climate.nasa.gov/news/2865/a-degree-of-concern-why-global-temperatures-matter/.

Net Zero Tracker. n.d. https://zerotracker.net/."Found: Closest Link to Eve, Our Universal Ancestor." New Scientist. www.newscientist.com/article/mg22429904-500-found-closest-link-to-eve-our-universal-ancestor/.

Ollerton, A. 2023. In *Romans: A Letter That Makes Sense of Life*. Hodder and Stoughton, eds.

Piketty, T. 2017. *Capital in the 21st Century*. Harvard University Press.

Poushter, J. et al. 2022. "Climate Change Remains Top Global Threat Across 19-Country Survey." *Pew Research Centre.*

Prowle, M.J. 2022. Reforming UK Public Policy Through Elected Regional Government. Routledge.

Raup, D. March 28, 1986. "Biological Extinction in Earth History." *Science.*

Rawarth. K. 2027. "About Doughnut Economics." https://doughnuteconomics.org/about-doughnut-economics.

Raworth, K. December 2017. "Why It's Time for Doughnut Economics." *Progressive Review.*

Ridhwan, M.M, Nijkamp, P, Ismail, A, Irsyad, L.M. 2022. "The Effect of Health on Economic Growth: A Meta-Regression Analysis." Empirical Economics 63, pp. 3211–3251, https://link.springer.com/article/10.1007/s00181-022-02226-4.

Schweickart, D. 2011. *After Capitalism*. Rowman & LittleField.

Stark, T, N. Osterberg-Kaufmann and S. Pickel. 2022. Dysfunctional Democracy(ies): Characteristics, Causes and Consequences. Zeitschrift für vergleichende Politikwissenschaft 16, pp. 185–197. https://doi.org/10.1007/

s12286-022-00537-5.

The Carbon Cycle. n.d. https://d32ogoqmya1dw8.cloudfront.net/images/eslabs/carbon/global_carbon_cycle.png

Tongia, R. 2021. "Net Zero Carbon Pledges Have Good Intentions. But They Are Not Enough ." www.brookings.edu/blog/planetpolicy/2021/10/25/net-zero-carbon-pledges-have-good-intentions-but-they-are-not-enough/.

University Of Cambridge. 2007. "New Research Confirms 'Out of Africa' Theory of Human Evolution." ScienceDaily. www.sciencedaily.com/releases/2007/05/070509161829.htm.

Vaughan, A. November 1, 2014. "Recycling Rates in England Have Stalled." *The Guardian*.

Warner, J. October 8, 2022. "Few Have Yet Woken up to the Full Horror of Britain's Diminished Situation." Telegraph.

Watkins, E. 2018. "GOP Congressman Asks If Rocks Are Causing Sea Levels to Rise." *CNN*. 18 May 201. https://edition.cnn.com/2018/05/17/politics/mo-brooks-nasa-climate-change/index.html.

WEF.2021(b). What's the difference between 1.5 and 2 degrees of global warming? World Economic Forum. https://www.weforum.org/agenda/2021/07/2c-global-warming-difference-explained/

Willage, A. 2022. "The Stark Difference Between Global Warming of 1.5°C and 2.0°C. *Forbes*. www.forbes.com/sites/mitsubishiheavyindustries/2022/01/26/the-stark-difference-between-global-warming-of-15c-and-20c/.

Williams, M. 2006. "Deforesting the Earth: From Prehistory to Global Crisis." University of Chicago Press.

WSJ. March 24, 2023. "The Real Cost of the Inflation Reduction Act Subsidies: $1.2 Trillion." *Wall Street Journal*.

About the Author

Professor Malcolm Prowle originally trained as a scientist but switched to a career in economics and finance. He has gained extensive experience in both the public and private sectors and has advised government ministers, ambassadors, and senior public servants on many public policy issues. He has served as an adviser to two House of Commons Select Committees and three shadow government ministers. He has also consulted for several UN organizations. His experience in the fields of science, economics, and public policy makes him ideally suited to write this book.

Index

OTHER TITLES IN THE ENVIRONMENTAL AND SOCIAL SUSTAINABILITY FOR BUSINESS ADVANTAGE COLLECTION

Robert Sroufe, Duquesne University, Editor

- *Making the Connection* by Peter Sammons
- *Sustainable Investing* by Kylelane Purcell and Vivari Ben
- *Confronting the Storm* by David Ross
- *Sustainability for Retail* by Vilma Barr and Ken Nisch
- *People, Planet, Profit* by Kit Oung
- *Bringing Sustainability to the Ground Level* by Susan J. Gilbertz and Damon M. Hall
- *Handbook of Sustainable Development* by Radha R. Sharma
- *Community Engagement and Investment* by Alan S. Gutterman
- *Sustainability Standards and Instruments* by Alan Gutterman
- *Strategic Planning for Sustainability* by Alan S. Gutterman
- *Sustainability Reporting and Communications* by Alan S. Gutterman
- *Sustainability Leader in a Green Business Era* by Amr E. Sukkar
- *Managing Sustainability* by John Friedman
- *Human Resource Management for Organizational Sustainability* by Radha R. Sharma

Concise and Applied Business Books

The Collection listed above is one of 30 business subject collections that Business Expert Press has grown to make BEP a premiere publisher of print and digital books. Our concise and applied books are for...

- Professionals and Practitioners
- Faculty who adopt our books for courses
- Librarians who know that BEP's Digital Libraries are a unique way to offer students ebooks to download, not restricted with any digital rights management
- Executive Training Course Leaders
- Business Seminar Organizers

Business Expert Press books are for anyone who needs to dig deeper on business ideas, goals, and solutions to everyday problems. Whether one print book, one ebook, or buying a digital library of 110 ebooks, we remain the affordable and smart way to be business smart. For more information, please visit www.businessexpertpress.com, or contact sales@businessexpertpress.com.

www.ingramcontent.com/pod-product-compliance
Lightning Source LLC
LaVergne TN
LVHW020847190325
806302LV00007B/97

* 9 7 8 1 6 3 7 4 2 7 2 4 8 *